BIGFOOT
ADVENTURES

GARY AND WENDY SWANSON

Cover photo was sworn to us as a shot taken by an amateur photographer. Two seconds later, the Sasquatch disappeared behind the tree, and Jerry said he heard it crashing through the brush!

Hidden in the cover photo is what we feel for certain is a Sasquatch, although not clear enough to be proven. Look at the area right above the small waterfall.

Unless otherwise credited, all images were provided from the files of Gary Swanson.

Published by Swanson Literary Group
ISBN: 9781087442723

Other books by the authors:

CONTENTS

vi

INTRODUCTION

After retiring in southern Oregon, we began hiking through the spectacular scenery and taking photos for later reminiscences and records for return trips. That led us to introducing a unique guide to our favorite hikes with a link to over 1,800 online color photos. That book is entitled "Hiking Sasquatch Country" and it contains commentary as well.

Not too long after starting this project, we had our first encounter with Sasquatch! Our next surprise came when due to our amazement at this event, we began talking to other hikers we met on the trails about our venture, and that's when we learned just how very many people out there who have had sightings and encounters, but who seldom talk about their experiences for fear of others questioning their sanity!

That started a series of published reports from people who began sending us their accounts of their personal experiences with the Bigfoot/Sasquatch, with the promise of confidentiality.

We are pleased to announce our latest collection of personal experiences by those who have encountered Sasquatch. They shed new light on the subject of hibernation, and we are becoming even more certain that the Bigfoot likely does so in colder climates, but perhaps not in warmer parts of the country with more mild winters.

The majority of the encounters in this book come from the Pacific Northwest, but there are also stories from Montana, Minnesota and British Columbia. We take them in order of receipt out of fairness.

Many gold miners, hikers, hunters and those living on rural properties have had experiences and sightings with these elusive creatures, and they want to share their stories with you.

We received the story of one encounter from an undercover submitter of his personal experience on a remote logging camp that may indicate a mass murder of many resident Sasquatch, but it is sketchy, and as yet, no actual evidence has been produced, and no further proof would make it less disturbing.

If you have had a personal encounter or sighting of a Sasquatch that you would like to see published in our next book, please send the details and photos to:

swanliterary@gmail.com

If your story is published you will receive a copy of the book as our thanks.

1 AIRBORNE BIGFOOT DIDN'T LEAVE TRACKS

Jonathan ~ British Columbia

I'm submitting this story through a U.S. subsidiary of my father's company, as he has extensive interests in a newspaper publishing chain across Canada with many investments in the USA. I wish to keep our family name out of the records; primarily, because it sounds insane! Although I lived it, I still have many questions.

It happened on a British Columbia bear hunt. My father and a small group of his major investors and top company bigwigs assembled a varied group for an annual bear hunt near the Cutbanks in Prince George. I don't hunt, as I am against killing animals for sport, however I had recently joined my father's newspaper chain and I was in the process of training to become a wildlife photographer to work along with their various publications and magazines.

Father thought some outdoor experience would broaden my abilities and further enhance my reactive senses for those rare opportunities when seconds make the difference between a great photo and a "click."

I was glad for the opportunity to share an adventure such as this, and for the chance to experiment with my classroom study and to fail or succeed with only myself as the judge of my progress.

After setting up camp, our camp cooks and an entire staff of people were there to assist with the hunters' (myself included) every needs. Such extravagance could only be likened to a vacation at a luxury resort; the only reality being the periodic snow, and the luxury suites being tents with propane space heaters and several portable and relatively crude latrines.

Anyway, I quickly fell into the established routine and I began to experiment with short trips out of the main camp. I was pretty much on my own and my only real duty was to extensively photograph every elk and bear along with the hunter who bagged them. This allowed me to pretty much do as I pleased, especially in the mornings.

The only places that had snow this early were at slightly higher altitudes, but the hunters had to bring their trophies down to the camp, so I was lucky.

Nechako River at the Cutbanks in Prince George, British Columbia

I was into a routine by the second day that involved walking along a marshy area on a trail on the hillside that was well used by local game. Deer and elk hoof prints were plentiful, as well as fox and possibly wolf prints. These were much larger than the fox prints, so I made the assumption.

Then, on one of my trips I was off on a side trail when up ahead I saw what I thought was a bear, and my inexperienced brain didn't register anything wrong until having watched it for a few minutes as it was reaching up and bending down a branch on a tree, and then it continued on.

It slowly walked in a slouching manner, somewhat as the gorillas and apes I had seen in zoos and circuses. Then, as though something had frightened it, the animal suddenly reached above its head, grabbed a tree branch, and in a few seconds, it had gone swinging from pine tree to pine tree until I could no longer see it.* All of this without a sound and traversing through this thick forest as though leisurely swinging over the "ol' swimming hole!"

I was really shocked, and that evening in the mess tent, I cautiously dared to tell my tale at the first pause in conversation. I was hesitant to even relate what I had seen, but I began by first prefacing my story with, "Now don't laugh, but I think I saw a large ape today." No one laughed, and I wondered if they had all gone deaf, but as I continued, I began to see curiosity rather than the ridicule I was expecting. So that that's when I stopped and looked around the table and asked why nobody was laughing or calling me nuts. That's when I was told simultaneously by about six people; "You saw a Sasquatch!"

For the next hour, I heard story after story of other people's experiences with these Bigfoot animals of which I was so unfamiliar. The one part of my story though that drew argument is when I said the big creature took off swinging hand over hand through the trees. That had everybody taken aback, because they had never experienced seeing one of them do that.

I received the most criticism however, from everyone after one of the executives held up his hands for silence and looked me straight in the eyes and asked why I was on this trip to begin with. His loud, critical demand suddenly shocked me into reality, when I realized at the same time as everyone else did; I was the one person in the whole group who should have snapped a photo! Then one of the men asked if this was proper procedure for training wildlife photographers. The roaring laughter seemed much louder, and I'm certain it was because I was the boss's son!

Well, I was never allowed to forget that by anyone in the camp! Every day when the hunters left camp, at least one of them would loudly say something along the lines of, "I wonder, since I'm a hunter, should I carry some cartridges or should I not?" Then everyone in the whole camp would roar with laughter! I knew that I deserved every bit of it, and I'll never be without a camera again.

I made daily trips around the area, but I never saw the Sasquatch again. I really did see him do a "Tarzan act" through the trees though!

Publishers note: We have received numerous reports of Sasquatch using this method of travel in heavily forested regions.

2 WE PROTECT OUR SASQUATCH NEIGHBORS

Anonymous ~ Western Montana

When I was growing up in western Montana (I cannot identify the exact location, because my family has eight generations of settlers here), we had a lot of nearby relatives. Going to school was quite different from what I assumed other schools to be, as I was related in one way or another to over half of our class of 46 students!

Our beautiful valley among the mountains wasn't isolated from the world as it might seem from my description, but we did not have close ties with what we arrogantly referred to as the "outside world." The difference that I have finally decided to adress is the fact that we all pretty much have kept a secret within our community that has been carried through many generations; Sasquatch!

In today's highly enlightened world, this mysterious creature does not stir up such an excitement as it would have even such a short time ago as 10 years back. Reason being; it goes back to how we all grew up. Among our residents there has always been a comradery of isolation from the outside, and an independence and a certain smugness that is sort of like some of the coal towns of the east (from what I have read).

Our "living" came from a mutual effort centering entirely around agriculture and with raising livestock as the core of our existence; we all earned our livings as a self-supporting community of like-minded individuals. The only thing most of us needed from other people were the sources to purchase the things that they manufactured.

www.needpix.com/photo/522111

We made our own money and we always were taught by our parents that the only thing we needed outsiders for were things to make our lives more comfortable, and to buy what we raised for sale.

A great part of our so-called arrogance stems from the fact that we also had an in-bred desire to maintain our secret of having our own colony of Sasquatch! Although nothing is in writing anywhere, our community is well aware and concerned for the well-being and privacy of several small groups of these Bigfoot people that live in a secluded series of beautiful mountain valleys, isolated to all but our private roads.

They have, for several generations, been allowed to help themselves to food crops and the occasional calf, cow or sheep to supplement their preferred subsistence on wild game. This sharing procedure perhaps harkens back to the first pioneers who are our direct forebearers who allowed the local American Indian tribes to occasionally take a cow or two for their use during tough times.

We have a community wide unspoken, but assumed agreement to never disclose our Bigfoot neighbors to outsiders, and the few times it has happened have resulted in penalties enacted on the perpetrators so severe that no one can ever know!

Please understand that we may be a little harsh about protecting our own, but it is a necessity and we are "dead serious!" I guess if you really came to know us you would understand. We, however, won't allow you the chance!

I know that in all likelihood you won't want stories that aren't exciting, but I wanted to confirm that the animals are real.

3 SASQUATCH IS AN OMNIVORE

Tommy C. ~ Central Washington State

Publishers note: A submitter who has contributed a story in one of our other books has responded to the often-asked question as to whether the Sasquatch is a carnivore, and if so, why do we not find more evidence of this fact?

We'll turn it now to our friend and contributor, Tommy C.

I have heard a lot of questions about what the "big fellows" eat. Many people who have seen the Bigfoot as many times as I have over my life on the family property for 67 years now, should realize that the big guys are not the ravenous eating machines that you might expect.

First of all, for those of you who have seen it run may have noticed how much its hair fluffs in the breeze. If you think about how small its body is beneath all the wispy, long hair all over its body, there's not a whole lot of animal underneath!

At times on our huge property I have seen them when they didn't know I was there (not easy to do) and their fur seems to be made up of very fine hairs that I have examined when I found sheds that caught on berry bushes where they had been, and they seem to be covered in very long, fine hair that I suspect to be made up of a sort of a hollow center. I've been at times between the Bigfoot and the sun, and at certain times, I caught glimpses that gave me the impression of tall, gangly bodies that almost seemed "lizard-like" in my memory.

Judging this to be true, there really doesn't seem to be much body at all to feed. Maybe that's why we have so little evidence of freshly killed animals, bone piles or missing livestock; certainly there are some, but not what one might expect from a large, ravenous carnivore!

In addition to this, I can assure you that the Bigfoot is an omnivore*. I know for a fact that over all of these years of having a year-round family of them on our back property, we have learned to live together.

For several miles all around our land, there are "peat bogs." If you are not familiar with them, you may liken the land to perhaps a Florida swamp in the fact that trying to walk across it takes you from one moveable hummock to the next. It's maybe similar to a football field covered about three feet high in pillows of various sizes, so every step you take, your body slumps downward as you go one leg to the other. Plus, there are deeper holes every few yards where you sink into water that gives no indication of any solid bottom, so you just keep going and going. The only places safe to stop and rest are the larger hummocks that may have a small tree and maybe some cranberry bushes.

Traveling through one of these areas from higher spots to small patches of trees, and over miles where the largest tree may be 20 feet high with very few and very high branches and not a single patch of gravel, or even a rock, makes one constantly aware that there may be no solid ground anywhere! The only trees are an occasional Tamarack. In winter, it's even more eerie, as Tamarack's needles die and turn brown.

Having lived my entire life on the family homestead, I have learned how to cross these marshy areas, and in many ways it's comforting to live on our many acres of "high ground" situated in the midst of this massive private area.

What none of our neighbors have ever known is that far behind our homestead, in the center of the vast hummock covered acres and the forests of dead and dying poplars, is over 120 acres of higher ground that connects with thousands of acres of basic wilderness that crosses into Canada. On this higher ground lives the nicest family of neighbors one could ever have.

I have never had the stamina to venture that far back, and even if I had, I would not wish to disturb our furry neighbors, but we know they visit our acreage, and every year we plant extra acres of corn, beans and potatoes, knowing they will visit us throughout the harvest! We pick by day, they visit by night.

We also lose chickens, turkeys and lambs periodically, and when we butcher the occasional cow, we make a point to hang a share on a hook outside the barn high enough so they can reach it, but other animals can't.

Our dogs surprised us by the fact they do not ever bother the Sasquatch, and outside of an occasional bark to alert us, they allow the big guys to come and go undisturbed. I guess they learned while growing up, the rule of coexistence.

The only thing I can't be certain of is whether or not the Sasquatch hibernates. Their visits come less frequently once the snowfall gets deep. My son has flown over the area behind us where the Sasquatch are assumed to live and he circled around the area, and although he could see trails in the snow, there was little sign of any big amount of movement.

He did notice, as did his co-pilot, wherever there were downed trees, there looked like pine boughs piled against and over them with a lot more snow over everything than what would have fallen naturally. This does give rise to speculation that the group maybe does hibernate, and they may also dig way under those logs as well.

Also, there are records that show that area back there had at one time been the source of a brief period of mining, as there had been a pocket of anthracite coal that had been too small for a profitable venture, and it was abandoned as it was, so there could still be some open shafts that the Sasquatch could shelter in which adds to the hibernation theory. Since their metabolism would slow way down, that would explain the fact that they don't forage much after the first snow.

Publishers note: Wikipedia.org defines the word omnivore in this manner. "Omnivore is a consumption classification for animals that have the capability to obtain chemical energy and nutrients from materials originating from plant and animal origin. Often, omnivores also have the ability to incorporate food sources such as algae, fungi, and bacteria into their diet as well.

4 GOLD FEVER BIGFOOT

R. B. ~ Brookings, Oregon

I had an experience last spring and I finally dare talk about it after I happily found your website. Please accept this account of what happened to me.

My hobby now that I retired to the Oregon coast has been searching for the treasure that made Oregon and California the destination for early settlers by the thousands; GOLD!

Cacophony [CC BY-SA 3.0 (https://creativecommons.org/licenses/by-sa/3.0)]

We get a lot of rain here, but the temperatures are okay and the scenery is spectacular! I began hearing tales about Bigfoot as soon as we were settled in our new home; and no less than a dozen of our new acquaintances have told us of their personal encounters with this fascinating, but reclusive creature. One couple's encounter is in one of your books, and they convinced me to share mine.

I was cautioned early on to be on the alert, because my new hobby was to take me into Oregon's coastal mountain range in my search for gold. I won't say where this incident took place, except that it was within 50 miles of where the Rogue River enters the Pacific Ocean.

When I began my new hobby, I really got lucky. Being fortunate to befriend a new neighbor with similar interests (a retired geologist), we hit it off immediately, and thanks to him, within a relatively short time I had found "color!" There it was; a few tiny flecks of gold in my pan; that's all it took to hook me! I have since then, made a practice of searching different areas and anywhere there are trickles of water to wash pans of gravel.

That's where I first saw the elusive and very frightening beast called Sasquatch; which when I saw the first one, I knew immediately what it was, but even so, I was scared! This huge beast was probably over eight feet tall, but lurched through the brush like a tank after it stumbled on me squatting over a small stream.

When our eyes met, it gave out sort of an annoyed "huff" and it was gone in seconds! The small, but noisy stream drowned out any chance of hearing where it went from where we met, but I secretly hoped that it didn't mind sharing this area with me.

Since my first encounter with the Bigfoot, I spent three more trips to this spot before I saw it again; only this time I noticed it out of the corner of my eye. It was just sitting on a large, flat rock, and it was only after I had passed by that I realized it was there. As I turned, we made eye contact; it made kind of a snort and slipped around the rock, and I heard a few branches crack in the patch of woods, and then it was gone!

Over the last six months I only saw it twice more, and after all this time, I still don't have enough gold to bother going in to have it weighed or assayed; however, the thrill is still worth it!

I wanted to at least tell my Sasquatch story, even as boring as it turned out. I can testify that the elusive animal is real! It was about eight to nine feet tall, walked very similar to what an ape would, but not nearly as slouched over; its stride is more upright like a human. Its arms don't swing like the gorilla, and its paws are more like large, human hands, but the fingers seem longer and the palms longer also.

My impression of this animal is that it seems far removed from the slouched and stooped gorilla, and everything about it tells me it is most similar to a human giant such as I once saw on television. The feet seem longer, but the toes struck me as being on the short side, and I did not recall any sign of claws, either on hands or feet.

It didn't have the flat, pug nose that gorillas have, and yet I couldn't really say that its nose stood out as much as a human nose either. Again, its gait was hard to determine, as most every encounter I had with it, we were both moving, and not being a trained observer, I have to rely on my "impressionistic memory."

Thinking about its ears, I recall them being small, but also covered by longer hair. The hair all over its body was unlike that of a gorilla, in the fact it seemed longer and fluffier; like I have inside a winter parka I bought at an outdoor store in Canada. The nap looked similar.

I think I'm accurate in guessing that I saw three or maybe four different Sasquatch, and there was a difference in my perception of their eye color. It may have been largely due to a difference in lighting, as the sun is often a stranger near the Oregon coast, and somehow, I am left with the impression of a range of eye color from a darker reddish tinge to a sort of gray and yellow combination.

Once, I saw the biggest one staring at me, and its eyes seemed like they were flickering sort of bright red and yellow, and then it began to move and the eyes turned a deep burgundy shade, and then instantly changed to a cold, black color as I turned and "beat feet!"

Another time, I found what I assumed was a small patch of Sasquatch fur that had snagged on a patch of mountain blackberries, and it almost seemed like the fur I get from brushing our golden retriever Sally, except it was thicker and a very dark color, but the individual strands of hair were thick like the diameter of a large sewing needle. The stuff had an odor like a wet dog, but allowing for where I found it, who can tell?

The winter is on its way now, so I'll try to give you a follow-up next year, but my "gold fever" has pretty much subsided now, so I may just concentrate on taking a camera and going on a Sasquatch study next year. Plus that, my wife has developed an interest in seeing this elusive being for herself! She can't see any sense in the backbreaking effort of gold panning, but the people in fur coats sound intriguing to her!

5 BACKTRACKING BIGFOOT

T.C. ~ Methow Valley, Washington

My wife and I live in a beautiful area of Central Washington. Our home sits on 35 acres with a picturesque forest behind us that goes all the way to a series of hills that gradually become a solid wall of rock. The sun sets just so we can enjoy it, but the canyon on the other side of our land gets dark two hours before the rest.

It's that darker side where a family of Sasquatch live. In my younger days, I quite often hunted deer back in the series of canyons, and way off, there is another mountain about 1000 feet high with a series of sharp, steep cliffs.

360° panorama near the summit of Goat Peak in Okanogan National Forest of the northern Cascade Range
By Farwestern Photo by Gregg M. Erickson - Own work, CC BY 3.0,
https://commons.wikimedia.org/w/index.php?curid=7781996

What lies next to our property line is all government land, so unless someone gets our permission to gross the gully on our private bridge, there is no access to the area at all. I only mention this because years ago, I placed a sign that the bridge is unsafe and chained it off due to deteriorating planks that I no longer had any reason to replace.

Seldom does anyone make this "out of the way" drive anymore; the reason I know is because, last month I decided to check things out over on the other side, so I climbed on my new 4-wheeler ATV and with my revolver and a lunch, I was back in my old haunts in a half hour. Wish I'd have bought one of these little 4-wheelers years ago!

I curved around a forty foot high berm that acts as a natural shelter that deflects winds upward and leaves the most beautiful sandy area that is lightly covered with grasses and small trees. Leaving my ATV, I embarked toward the distant hills that eventually grow upward in varying stages to the steep mountains about two miles distant. These foothills were my favorite haunts until I broke my hip five years before, and now I relived old memories as I walked.

Being alert for signs of the Sasquatch family that I assumed were still in these switchback canyons, I soon found a large footprint where a small rivulet coursed away from the stream that meandered alongside my path. Stopping to examine the print, it was the same type as I remembered, and the creature that left it had obviously no fear of being discovered, as there were more signs of its tracks as I continued. Where the trail was dry, every so often I could see impressions of toes in the soft, drifting sand of the trail.

I was nearing what I had nicknamed "the maze," and the circuitous route I had to now navigate was a series of sand dunes that over centuries, had accumulated from the steady winds crossing the tops of the high buttes that rose above this canyon. It had remained untouched by humans other than my wife and I and our two children for the fifty-odd years we have owned it; I felt like a stranger on my own land!

I had noticed as I walked that several times there would appear a quick cloud of dust ahead and alongside my path at various intervals, and I knew exactly what the cause was. My thoughts went racing back to my past experiences with the Sasquatch, and I wondered if they had a recollection of the times we briefly stared at each other over dunes and logs, or if they even had such ability to remember as we humans do.

After another curve in the path I came to a place where there were several tracks of very large, humanoid critters that could only be them! I had now continued more slowly, and the trail then split in opposite directions; almost at right angles. The tracks I was following appeared clearly as they went to the right. From the trail on my left, was another set of tracks that came down the hill and joined the ones I had been following; and now the tracks were leading me downward at a slight angle. There was a thrill building inside me; that feeling of excitement as I anticipated meeting two of the Sasquatch face to face, as I hadn't experienced for years!

Taking my time, and with periodic sips from my canteen, I barely noticed that one set of tracks had stopped; that is, until the sudden realization that the other set had also stopped. No way possible could this happen, as on one side of this three-foot wide trail was a high cliff, and on the other was a steep drop of at least 20 feet! As I retraced my steps, I was totally baffled and absolutely dumbfounded!

Once I had arrived at the place I had begun my downward trek, I was about to return to my rig, but something about the set of tracks coming down from the trail above was bothering me. As I studied them, then turned to the tracks that led downward where I had just followed, there was a difference! Still not being able to understand, I had to know!

Looking upward at the prints I became suspicious and decided to climb that trail, and as I reached the point where the trail had first appeared, I looked down, and suddenly noticed what was bothering me; the tracks I was looking down at now were deeper than the tracks leading down the other trail.

I quickly turned and went further up this new trail, and suddenly, the footprints had turned completely around and were now leading upwards. Climbing quickly, I soon reached the top of this ridge and I could see the tracks still proceeding, only now further apart as though the animal was now running. I had been duped!

Turning back around, I knew further pursuit would be futile; I had been tricked! Looking back to where I had followed the original tracks to the right and down, the second set that had appeared to be going in the same downhill direction and had then stopped; as had the ones I had been following were made by the same Sasquatch! The second set were made alongside the others by the same critter who stepped to the side, and then began walking backwards up to the point where the trail curved upwards. When the Sasquatch reached the main path, it continued walking backwards up the hill and around the corner; and then as I found, the critter turned back around and climbed to the top in its' normal manner.

I had never before heard that the Sasquatch were this intelligent, and this experience both impressed me greatly with the that fact and wore me out in return!

6 I'LL TAKE KING KONG OVER THE ARMY

Paul N. ~ Northern California Coast

I was visiting a friend whom I hadn't seen since high school. After graduation I went into the Army and somehow survived 20 years. I had planned to stay in for a while longer when an unexpected inheritance from an uncle whom I hardly knew, took all desire to work out of my system.

Returning to my birthplace in northern California, I found several of my childhood friends had remained in our small town, so I set out to renew old friendships. One of my best friends from school, whom I'll call Barry, had taken over his dad's business, which will remain unidentified; for soon to be discussed and obvious reasons!

Barry had a most leisurely lifestyle with his business, as it virtually ran itself, so with his employees, whom he paid very well, running the operation, Barry pursued the same hobby he had begun in his junior year in high school; gold mining.

After a few drinks one night, Barry invited me to visit what he called his "secret hideout." Now I never asked, nor did I ever find out if his gold mining operation was legal or not, as I really didn't then and still today, don't want to know! I reasoned, if it's legal, I'd wonder where the competition was, and if it's illegal, I don't want to form any judgements about an old and rekindled friendship.

Taking off for the "wilds" at three o'clock in the morning had been a challenge in itself, but I began to reawaken as the Jeep went airborne after a surprise bump in the ever worsening road. We didn't bother trying to carry on a conversation, as the soft top Jeep, combined with the noise of the rough road, and Barry's constantly fighting the steering wheel, weren't conducive to casual discussions.

Barry did mention at the start that the first part of our trip would be a lot of fast driving and rough roads, because he said we needed to get deep into the forest without being seen by any people camping in the area. Elaborating on his reasons, Barry explained that tourists will quit often avoid most of the more remote areas unless they see someone else go there, and then, they get the courage to follow. Followers, he explained; we didn't want!

The roads we were on steadily grew worse, and I hadn't even seen a tire track for two hours, and now, we were driving slowly on a narrow set of two ruts, down a rough, barely discernable pathway with the bumper mowing down a three foot high growth of grasses and weeds growing in the center. After about an hour of this, the road came to a fairly circular turnaround. On the right was a small pond about a hundred feet across, and a creek flowed in to it at one end, and then I could see where it exited on the other end, a ways across from where we were.

Then, just as I was prepared to step out, Barry turned hard left, and with the front bumper pushing tall grasses down in front of it, we came out on a fairly open, rocky road. To my surprise, Barry turned off the ignition, reached in back and withdrew a heavy duty push broom and made his way back down our trail to a ways before we arrived at the wide spot. He said the words, "For our privacy," as he labored with an obviously well-practiced routine of brushing out our tire tracks and pushing the grasses that our tires had flattened, to a more upright position.

Satisfied that we had arrived unseen, Barry carefully scanned the horizon all around, which he said he was checking for signs of campfires. Then we went back to the Jeep and were off once more. Another hundred feet and we were stopped by another gate that looked like it hadn't been opened in years, but stopping in front of the old relic, Barry was out in an instant and I saw him withdraw a key from his pocket and open the ancient looking lock. We drove through the gate and after quickly re-locking it, we repeated the act of straightening grasses and clearing away tire signs, and now we were off into a fairly open valley amidst a wilderness area such as I had never seen! It reminded me of my first landing in "Nam," although nobody was shooting at me this time.

After about 20 minutes of uneventful driving over a rutted and narrow trail, almost completely devoid of grass, Barry made a sharp right turn and parked under a small grove of fairly tall and wide pine trees. As he quickly bailed out, saying, "We're home," I followed his whirlwind system of unloading, and in what seemed like only minutes, the tent was up and Barry was setting up the small table, while I popped open the camp chairs. Then Barry stood up, grabbed a backpack and said, "Let's go to work!"

I then learned how to pan gold! This was the first time I had ever even seen any gold in the raw, and my jaw must have hung open in awe as Barry was laboring steady filling and working over his portable sluice box; this was crazy! As I said, I had never seen raw gold, much less be bringing shovels full of gravel and feeding a sluice box!

Barry's hands seemed well-practiced as he used a small, stiff brush to spread and separate the shovels full of gravel and rocks while I kept up a regular flow of water into the sluice by a manual pump to wash out the dirt while we both kept a keen eye out for that shiny, yellow prize money! Not being used to working so hard, I soon tired and was extremely glad when Barry called for a break.

The next part of our adventure was right up my alley; sitting in a camp chair drinking a beer with one hand and holding a stick attached to a hotdog with the other. Barry explained that the hard work of the last hour and a half was just to show me quickly what it was all about. He said that he normally takes it a lot easier, and then he showed me the glass tube of gold pieces and flecks, and explained that in a bit over an hour we had earned about two hundred dollars! Plus that, he explained further that most any decent-sized nugget was worth many times its' weight, and he had a ready market for nuggets from a buyer in San Francisco.

That evening, still relaxing in front of the campfire, the sound of an airplane engine reached us, and in a flash, Barry had flooded the fire, and then in only a couple more seconds, he had thrown a large camouflage tarpaulin over the Jeep and motioned for me to join him under the pines. The plane appeared to be about a mile away, and it seemed to be following the same stream we were on as it wound its way through the short, rugged terrain.

Barry laughed at the surprised look on my face, because when he burst out and pointed at me, I realized that my mouth had likely been open as I wondered at all the secrecy. Then Barry said that he should have warned me, but he had forgotten since we were such good friends; he said he just thought I knew.

Then I learned that the reason for all the elaborate secrecy was not because of keeping people from finding his secret gold mine at all; it was because this was not only the property of the United States government, it was within an area of land, that on the far end of this extensive row of mountainous areas, connected to a top secret Naval facility about 40 miles south of where we were camped.

Barry told me that years before, he had been shown this spot by an old Navy pal, and since they no longer checked the area by road, Barry had cut off their old padlock and replaced it with a similar looking one of his own. This way, he said if by chance they ever did check, they would think they had grabbed the wrong key and it wouldn't be a cause for concern.

That had been five years ago, and even their aerial reconnaissance was very seldom in the area as it was just too rugged and remote to attract visitors.

This was like Barry's own private gold-laden stream, and I hadn't felt such excitement for years! We rekindled the fire, and as the sun began approaching the horizon, we were still drinking beer and had finished eating a meal of bratwurst and baked potatoes when Barry suddenly put his forefinger to his mouth and quietly said for me to stay still; and with his left hand, he slowly reached for the remaining bratwurst, that when I had previously asked why the third one on the separate stick, he had simply replied, "For guests."

My thoughts ranged from maybe a stray dog or some other meat eating animal that he may have been visited by before, but other than maybe a fox, I couldn't think of a single animal that the brat would have been for, and since they were so huge, I couldn't conceive of seconds for either one of us.

As Barry slowly extended the stick holding the bratwurst to behind my left shoulder, I very slowly and hopefully imperceptibly turned my head a hair at a time until I plainly saw a huge, furry arm with a hand at the end of it, the likes of which I had never imagined could exist on an animal, and my thoughts envisioned "King Kong" about ready to rip my head off!

The hand however, had only one goal; and as it withdrew, holding the brat I turned clumsily to my left, and facing me directly on, was a monstrously huge ape face, but void of hair as would be on an ape. The animal, I knew immediately to be a Sasquatch! Even though I had never seen one before in my life, the animals were commonly known to live among our people in northern California. I had always heard stories growing up about Bigfoot sightings and regular encounters with our neighbors in Oregon and Washington, and up further north in British Columbia.

As I was receiving a return stare from this monstrous creature, I was amazingly calm, when my first reaction was to almost pee my pants. Now, as the giant animal quickly downed his snack, I saw his gaze turn to the bratwurst I was holding in my hand; about half eaten. Quickly accepting the fact that it must be safe, as Barry hadn't panicked like I would have if I had been alone, I shakily proffered my remaining half by holding it aloft, and I was amazed at how quickly the giant reached out, grabbed it, and then tossing the bun, it took about two bites and it was gone. I noticed a slight wince in its eyes, and realized quickly afterward that this may have been its first encounter with mustard!

In retrospect, as we discussed this incident, I mentioned to Barry that I hoped the poor guy wouldn't develop a taste for seasonings, because it would get messy around the campers in our forests!

7 SASQUATCH FRIENDS

Darla W. ~ Yreka, California

Our story is one that developed so slowly that we became involved with Sasquatch without even realizing it. My husband David and I are fortunate to have been blessed with wealthy parents. This fact allowed us to adopt a special needs child who requires a tremendous amount of care.

Since we couldn't have children of our own, we adopted this cute nine year old boy, and shortly after we purchased a fairly secluded estate with large acreage in northern California. This gave Jeffy plenty of room to explore and play.

Jeff was not mentally handicapped, but due to a horrific shock when he was three years of age, he had never learned to speak. Everything else about him seemed pretty much like any other nine year old; he could understand speech, but he had some sort of mental block against learning or even caring to communicate.

We have a service dog; a golden Labrador retriever named Penny, as that matched her copper-colored fur. The two of them went everywhere around our property and we finally learned to relax and let Jeff and Penny enjoy their time together.

We had about six acres and it had a four-foot fence surrounding all but the heavily forested far end that bordered national forest land that had its own chain link fence about six feet high.

Jeff and Penny began spending more time on their walks, and David and I were really pleased with the fact that Jeff seemed so happy here.

Then one day, David mentioned something that I had subconsciously been aware of, but that gradually became a norm that I just accepted without question. Every day when Jeff and Penny left for their almost daily excursions, Jeff grabbed his small backpack; a water bag placed inside and a collapsible plastic dog dish for Penny to drink from. He would maybe add some cookies or other snacks. I had become so accustomed to his independence, of which made me proud, that I loved to see such improvement.

David had noticed that lately, the backpack seemed to have been growing in size, so while Jeff was in the bathroom getting ready to leave, I sneaked a quick peek into the heavier than I remembered pack. Inside was the water bag, some cookies in a plastic bag and six hot dogs. I buckled it back up, and David and I were relaxing on our back porch when Jeff and Penny walked out with a "wave and a wag."

Three hours later we saw them returning and we just remained doing projects that we were engaged in. As usual, Jeff placed his pack on the kitchen counter, and with a hug and a kiss for Mom and a wave to Dad, Jeff went in to clean up while Penny was at her food dish, which I had filled for her as usual. David and I examined Jeff's backpack; he had already thrown the plastic bags away, so we checked the trash can under the sink, and the goodies were gone as we expected.

Somehow, as I made the grocery list each week, I must have become accustomed to replacing the supply of wieners subconsciously. When I checked back over several of the most recent weekly grocery receipts, I found that I had been adding hot dogs every week, even though thinking back; we had not had them as a family for over a month. David and I assumed that Jeffy was taking them as special treats for Penny, and when we discussed it, it made sense.

That evening as we were having dinner, I casually mentioned that fact that I had been restocking hot dogs an awful lot lately; Jeffy turned almost white and he hung his head, and I quickly asked him if he maybe had been giving Penny some "special treats," and he just nodded and looked scared, so I quickly said that it was a good idea which perked him up and he returned to his normal self.

Then I asked him if he could think of anything else she might like to have, and suddenly, he jumped up from his chair and ran to the pantry, flung open the door and pointed to the empty space where I normally kept breadsticks. I had totally missed the fact that where normally there were several packages the area was bare! When I asked Jeff which kind he liked best, he nodded to "unsalted." Suddenly my list grew as I'd mention an item and Jeff would either nod or shake his head, and my list now included apples and bananas.

Later that night when David and I were discussing Jeff's strange list, we wondered at many of the items that were included that we were sure that neither Jeff nor Penny were likely to eat, but the next morning we filled that list with Jeff's help. As we walked down the aisles in the grocery story, we also ended up with sweet corn and extra tomatoes.

We didn't try to question some of Jeff's choices, and we kept winking at this growing list of goodies that our excited son was adding as we went. Jeff of course, didn't realize that this was all going too easy for him, and he innocently, without suspecting, had us now very concerned! Maybe there was some transient living back in the woods behind our property and had been asking for Jeff to bring him food, but to keep it secret? This, we were determined to find out!

The next day, Jeff's routine was much the same, and David and I pretended to be involved in projects, and we made certain to keep away from the kitchen around the time Jeff normally took his walks with Penny. Today was different, as Jeff came over and gave us hugs and pointed to the field to indicate that he was off on his hike. We didn't mention the fact that he was leaving an hour earlier than usual, or that his bulging backpack was so full he almost staggered with it.

We were pleased that Jeff hadn't noticed that David and I were both wearing our hiking clothes. We were both so heavily camouflaged we would blend in quite handily for our "spy mission." As a precaution, just in case this was a circumstance of some transient or person that was controlling our son, we each carried side arms. Neither of us had visited this more remote part of our property since we moved in; the people we paid to mow the field grass were the only ones to ever see the forested part of our land.

We stayed far back, and the only way we even knew where Jeff and Penny had gone was by the obvious mat where their footsteps had left the tall hay laying down, and as we followed, the grass was already drying and bending upwards. Another hour and the trail would be impossible to follow.

The good part of it was that they were staying only a couple of hundred feet in from the fence line, and all along our acreage, as we walked inside the property line about 30 feet there was a double line of mature trees staggered randomly along the fence. On the right side of us was all state forest, as we remembered it was the same at the far end of our land where we were headed for now. The difference in where we were going now was that we had paid to have the entire fence at the far end removed upon our move-in. After the last few years one couldn't tell where our land and the state forest separated except by the end post on the fence on the right side where we were approaching now.

There was a light wind blowing in our faces, which David pointed out would muffle our footsteps and our smell so Penny wouldn't be likely to catch our scent and come bounding up! She was too well trained to stray very far from Jeff, but he was becoming more independent every year; which was great.

The trees suddenly turned into a gigantic forest, and glancing to our right, David pointed at the corner post where our fence stopped. We now entered a totally dense forest where everything had been growing unchecked for centuries. David took the lead, and we cut over to our left about 20 feet or so, and he pointed out the barely noticeable signs where Jeff and Penny had walked. We now walked single file on an obvious animal trail of which there were to be found in every direction.

Then I heard a child's laugh, as in the same moment, David's hand was gently pulling down on my arm until we were kneeling on the soft, velvety moss beneath a stately pine tree that had to reach well over a hundred feet high. All around this, almost brushless forest of giant pines that were likely close to this same height during the civil war; this area was crisscrossed with nature's highway system.

As we cautiously crept forward, only a few feet at a time, we kept low and with occasional stops to relieve the terrible pain in our backs from stooping and straining, we followed the sounds of Jeffy's laughter and Penny's occasional bark.

Her happy yipping was apparent as we crept ever so slowly; then, there they were. Slowly peeking around opposite sides of the huge tree, we were sheltered by the family of vine maples that grew around the base of this behemoth.

In a small clearing was a sight neither of us will ever forget; our son was pulling a banana from his backpack with one hand and holding another one in his other hand. He offered them to a smallish animal that looked like a live teddy bear, while holding his other hand out to what we instantly knew to be an adult Sasquatch!

They must have all been concentrating so much on their activities that we had the rare experience of this amazing picture! I almost came apart then; first of all, here was our son, with a giant gorilla-sized animal, and another one about Jeff's same size; and playing like humans!

David and I both looked at each other in disbelief and just as we looked around once more, we were facing a very alert pair of eyes returning our gaze from the monster-sized Sasquatch! Apparently, the mother, because she was staring directly at our tree and then Penny came running to us with her tail beating back and forth in excitement.

We stepped out from the tree, but kept bent over in a subservient manner, which with Penny coming over wiggling her hips with tail wagging, seemed to have a calming effect on the moment. Jeff and the little Sasquatch were sharing a hug! Of all the amazing things; I began to tear up when our little boy looked at us and spoke his first word ever, he said, "Friend!" I was too dumbfounded to even speak as I turned into an emotional wreck!

Then, to add to our amazement, the adult Sasquatch made a sound; it may have been attempting to say "friend," but at the same moment, the animal reached out her huge paw, picked up the backpack in one hand and her infant in the other, and headed for the thick forest behind her.

The surprising thing was that she did not run; and just before disappearing into the dark area, she again made that same noise; and then David loudly said, "Friend!"

That happened over a month ago and although we have not seen any Sasquatch since, our son has added seven more words to his growing vocabulary; one being "Saska;" this is how he answers when we say "Sasquatch." Close enough for now, and we are so excited we had to report this. We also have left occasional gift baskets for Jeff's friends! They are usually emptied within two hours.

8 FINDING SASQUATCH
(Before it finds you)

Captain Lee ~ Washington Co., Oregon

Publishers note: We met this gentleman several times as we hiked through the Oregon Mountains from the Willamette Valley to the Pacific Ocean, and since we enjoyed his tales of modern mountaineering, we asked him to submit his feelings on "Squatchin'".

Willamette National Forest

Introducing; Captain Lee

I can truthfully conclude after many years of being among these creatures in the remote wilderness areas of Oregon's coastal mountains that the only successful way that a human can find Sasquatch is to; "Be there when one comes along!"

If you were expecting more, then I can give you my further thoughts on the subject, but believe me when I tell you, after several years of studying the Bigfoot animals, the only true way to find one, that all of us who have seen them can agree on is, purely by accident!

No amount of "baiting" has ever been found to work, and I have lots of friends who have tried. An acquaintance of mine once spent three nights in the deep woods with a cage trap containing two fat chickens, when finally, a pair of coyotes came in and swooped them up, leaving the trap still set.

I could tell you true yarns for hours, but let's consider what you're really up against, and there are not any known standards to use for reference. The only two certainties that I know of are:

1. In the state of Oregon, if it's not raining, it's going to!
2. Sasquatch is real!

Anything apart from those two is purely conjecture. Many people are terribly disappointed when they spend their hard-earned vacations camping in our mosquito covered wilderness, or suffer the insults of our ever changing weather to no avail. I can assure a great many of these folks that they may not have seen a Sasquatch, but in our more remote areas, I can say that I am fairly certain that Sasquatch saw them!

One must keep in mind that once we leave the last road and set foot on a trail in these vast forests, we're out of our element. In order that we may have a chance to encounter this elusive animal, we must enter into its home.

Compare it to the odds of a mouse making its way through your house or your yard without you taking notice; if you think you wouldn't see it, then please don't ever invite me over for dinner. Even though I joke about it, having spent a large portion of my life on the roiling ocean and the pitching of a ship in high seas while trying to avoid a nearby buoy or smaller craft, gives one the jitters for lack of a calmer word!

In many ways the constant awareness and glancing here and there while searching for Sasquatch can be as nerve racking, but my past life makes me possibly more alert; out of having to notice everything. The danger level may not be as difficult, however, how do we really know?

After several experiences with the Sasquatch, I cannot say I wasn't frightened. I cannot imagine what it would take to not be afraid when facing an eight or 10 foot gorilla-looking ape with hands that could hold a bowling ball, and weighing around twelve hundred pounds!

I don't mean to digress, but just thinking back, I can picture so vividly my first and last thoughts on each occasion; from being scared so bad I shook, to falling into a jello-like shakiness even a long time afterwards in just recalling the encounter!

Another aspect that doesn't help the Sasquatch hunter is we humans cannot adapt to the wilderness any more than a Sasquatch would be at home watching TV in our living room.

When I hear about some of the Sasquatch encounters by strangers and city dwellers that come strictly by accident, while many people I know have spent countless hours tramping through these forests in vain, I have to conclude that Sasquatch must certainly be a lot like humans in the fact that some of them are just smarter than others!

On those rare occasions where a person does come face to face with the Sasquatch, maybe the "big guy" is just having a bad day. Maybe, as the Sasquatch is walking through the forest, he's daydreaming about the cute little "Squatchita" in the next valley?

Even with all of the time my select group of "Sasquatch hunter" friends and I have devoted to non-weaponized hunting, they are just too smart!

I need to get back on track, as the publishers asked me to write a chapter, but if I keep on, they'll have a whole book!

As far as the odds for and against a person actually seeing or encountering a Sasquatch, I'd say 80 to one; that's on a good day! Why such bad odds; even for those serious people who follow the rules of stealth? Why, because we humans stink!

We smell of everything other than the forest we are entering to find an animal that lives every day of its life away from the smells that we bring. We smell of perfume, cologne, cigarettes, hair spray, deodorant, starch, bleach, our automobile exhaust, boot dressing; you name it. So, you see why our Sasquatch encounters must happen by accident?

My personal successes have generally occurred after I had been in the forest for a while. I do of course; take certain precautions, like doing all I can to avoid those smells I just mentioned. I also picked up a tip from a friend and fellow Sasquatch enthusiast who taught me to camouflage my odors by rubbing my legs and arms with pine needles, or the sap from broken off pine branches.

My wife still can't figure out why I get so much pitch and tree sap on myself after "one simple outing." I have relegated a certain group of clothing to these endeavors so we have a pact.

The one very worst offender I saved for last, because of its ability to kill any chance of making any contact with a Sasquatch, except maybe glimpsing one miles away and on a dead run. That taboo is cigarettes! If you smoke, just enjoy your hike and don't worry about being stealthy enough, because you won't see Sasquatch; ever! Fire is the animals' greatest fear; you won't get close!

Dogs also, instill fear in the Sasquatch, as dogs can be ruthless in following a scent trail. Sasquatch with young ones will quickly evacuate an area; where normally they would just step further into the brush until you pass.

I conclude by using my personal philosophy that I developed out of my years of living a large part of my life traveling through this beautiful scenery; enjoy the trip and keep in mind that you are likely being watched, so stay alert. I guess to sum it up; you'll need a fair amount of luck!

9   I DIDN'T KNOW THEY WERE SO BIG!

Dick S. ~ Carlton County, Minnesota

As I followed the rather grown-over animal trail alongside the small stream, I was watching carefully for the tracks of the coyote that had once again killed two of my folks' chickens the night before.   This was the third time in less than a month, and these sporadic raids had been happening more frequently this year than ever since my folks bought the home 12 years ago.

When my mother called to tell me about last night's visit, I decided to drive over to their place and spend a few days of my vacation to see if I could maybe set a couple of traps and hopefully catch this nuisancy predator.

I carried a pack with three traps and a snare, along with a couple of plastic containers of bait; and I was slowly walking the old creek trail that I had traveled so many times in my last four teenage years when I heard a loud splashing; like a waterfall.   I didn't remember any waterfall where the creek came down from the huge hill behind my folk's property, but I knew there was a tall hill; probably what many people would call a small mountain, that separated their acreage from that owned by the federal bureau of land management.   I doubt seriously that my dad had ever climbed to the upper part of the hill, because it was all rock, brush, deciduous trees and pine trees.   As for myself, I had hunted quite a bit and camped a few times off to the side of this looming promontory, but I had never been more than a few yards into the twisted alders, sumac, and whatever else grew in this jungle-like end of the property.

With my folks living at the end of the county line, the dirt road dead-ended just past their mailbox, so everything all the way from the swampy area where I now stood to the nearest section of road a mile away was mostly a watery, swampy home to snapping turtles, ducks and a few billion mosquitoes. I could already hear the constant drone of the gray hordes as I searched for the easiest way behind and around the eight or so foot high growth of sumac and raspberry bushes that so aggressively dominated the area.

My goal was to set my traps at a place where I knew the chicken thief would find it easy to not only smell the bait, but have a quick and easy pathway to it, so as to rush up and devour it without exercising too much caution. For that reason, I knew that everything had to work just right, because coyotes are way too intelligent to give a person a second chance to kill them! I wore leather gloves so as to not leave scent on my traps.

I fairly stumbled onto the harder ground that I remembered was here, and I had missed it before, because the grass was so high leading across the flat prairies around this area. Only after I stepped into the trees did the ground appear hard and rocky again. This would be a perfect place for my bait, and as there were plenty of places where the traps could be concealed so the coyote would not suspect.

As I stepped around a giant oak tree that I noticed still bore my initials, along with those of a past sweetheart now long forgotten, I heard that waterfall once more. I was still unable to wrap my head around any area where that small, winding creek could be falling off the hill far enough to make that much noise, so I took off my pack and laid my load of equipment alongside the oak tree that effectively blocked all but a few of the suns' rays.

Making my way toward the sound of the waterfall, I wasn't expecting any surprises, but out of habit, I was comforted by the presence of my grandfather's old H&R 22 revolver that had been my constant outdoor companion since he willed it to me four years before. Out of habit when entering a strange area, my hand automatically reached back on my right hip to make sure the hammer strap was snapped, and then I was parting the long willow branches that stung so terribly bad in the winter, and were so pliable in the heat of summer.

Then the splashing suddenly stopped! I was trying to comprehend what had happened. I had also stopped the second the sound of the water did, and now I carefully lowered the foot that had been mid-air when all other noise ceased, and at the same time, while barely allowing body movement, I crouched low into the thin, wispy grasses until I could kneel my shaking legs on the damp ground.

Still no sound came from where I had perceived there to be a picturesque waterfall, even though deep in my memory, I had no recollection of more than a casual stream flowing from the hills ahead. I realized that I had incorrectly assumed that somehow a water source from the small stream had cut over a bit to erode a new path for itself, so something was not right.

A couple of minutes, which seemed like a half hour, went by, and then the sound continued as before! Relieved as I was that whatever it was that had been splashing and maybe playing in the stream was not scared away, I now had mixed feelings about what I was so glad about being prematurely so relaxed!

The most dangerous animals in this area were black bears and cougars; neither of which I was equipped to argue with! The lightweight revolver on my hip may be a cap pistol for what it would do against a mama bear with cubs, or a nasty tempered cougar, maybe with a little kitten of her own. Either way, I knew I could not outfight or outrun such situations, and then the decision was made for me; all of a sudden I heard the most godawful loud roar that made me almost mess my pants and lose my lunch at the same instant.

I envisioned a monstrous lion, that at any second, was about to pounce on me from the brush beyond and shred me into a pile of quivering flesh! I couldn't move, even if I knew what to do, when a huge shadow loomed above my kneeling, shaking form, and I looked up to meet the glaring yellow eyes of King Kong's son! The monstrously large ape sort of slobbered and growled at me, and I thought I was about to be torn apart, when the giant just kind of gave out a loud snort, like maybe a sigh of relief. I could only interpret the big fellow's reaction to be one of relief. Maybe it thought that whatever I was could be a real danger to it?

I can laugh now as I think back and try to imagine the big fellow's relief, when instead of a dangerous adversary that could do it harm, it looked over at my small ass; and I can now see the picture of the giant ape standing there, slapping its knees and roaring with laughter!

Anyway, it ran away and I advised Dad to maybe strengthen the fencing somewhat. After my experience, Mom made the ruling; "No more chickens!" Dad though, uttered a thought that reminded me of his father who was a Baptist minister, "Well, I don't want to deprive the poor animal of its food source." Dad immediately afterward purchased more chickens than before and swore me to secrecy!

10 GOLD PANNING WITH BIGFOOT

John Thune ~ Roseburg, Oregon

I walked the rutted dirt road as it wound along the ridge above the offshoot of the main river. This tributary was not nearly as wide, nor as dangerous, but to my liking; there was plenty of easy access to deer trails networking between the hill and the water. Following the rather crude directions turned out to be easier than I had anticipated, because after walking for about an hour and a half on the rocky road that was grown up with grass and weeds, I caught the pleasant scent of wood smoke; which I knew must be a sign I was getting close to my destination.

About 600 feet further I came upon a "no trespassing – keep out" sign. I yelled out a "hello," and continued toward what I could see was a makeshift cabin of faded wood; grey against the green trees surrounding it. A man stepped out from behind a tree and with a big smile; he said, "'Bout time!" We shook hands like old friends even though we had only met once before, and that was two months ago.

His name is Ken Belltrane, and we had been introduced by a mutual friend. Our mutual friend was an old Coast Guard officer who knew that Ken owned and lived summers on a gold mine, and had recently lost his partner when the man simply disappeared from sight and left all of his belongings in his sleeping area of the mine shack.

That had been several months with no trace of the man! Our friend knew that I was recently retired from the Navy and needed something to occupy my time, so after pondering such a "way out" adventure for days, I got word to Ken that I'd like to partner up for a while.

As we entered the rather small and crudely constructed miner's shack, I noticed another door directly to my left to which my host said, "Mine entrance." That was convenient; we didn't have to go outdoors to go to work.

We had a cordial meeting, and got to know each other better. I was to share the housekeeping chores, which were minimal, and Kenny would work the hard rock claim. My mining job would consist of panning the river below for the entire length of his claim; on both sides and inclusive of the bottom of the river when the river's depth was low enough to sluice its gravel bottom. This during low water, proved to be super profitable!

I agreed to split my findings, and he was more than generous by agreeing to accept a share of only 20 percent of my take. Kenny would report the findings to the state as required and do all the required claim maintenance that was also subject to state law. Kenny had agreed to this arrangement, plus, during the winter months, I agreed to join him working the inside main adit for a fifty-fifty split. We signed our two written copies and we were now partners.

I had my own pick, shovel and gold pan, and a large selection of glass test tube-like vials for my treasure, so I had Kenny give me a "crash refresher course" on the best ways to conduct my panning. Kenny had said that he believed this ground had not been worked for maybe over a century due to it having a continuously renewed claim, but very low activity. I could envision a hundred years of gold flakes and nuggets just waiting to get scooped up in my shovel!

It took a week of trial and error as I learned where to toss the gravel I had already checked, so as not to throw my discarded rock and sand where I might find myself panning it again at a future time. I began by working steadily, but with almost no results, when finally, after another bunch of instructions from Kenny, I began finding some "color" at last!

By the end of my first week, Kenny complimented me by saying I had earned about ten whole dollars! This left me eight dollars for my share. Eight dollars for a weeks' work, minus my share of the food! Kenny said I would get better, but he reminded me that I still had my monthly military retirement check coming, so I was way ahead of a lot of the hard working people along this river and in the myriad of canyons branching out around us.

As what became my usual regimen, I began taking an early morning jaunt along the cold, rushing river, and on my return one day, I was still about two hundred yards upriver from the claim when I spotted what in my mind was a "claim jumper!" There up ahead, was a man in a long, fur coat; bending over, with what I was certain was my gold pan, and judging how long he stayed bent over, he just may have found the gold that I had cleared the way for. I was really mad. Reaching in my coat for my constant companion, I removed my Smith and Wesson .38 Special revolver that I always had with me; as I was expecting that anyone so brazen as to work another man's claim within 50 feet of the claim sign was likely to be armed as well!

Then, as luck would have it, my foot slipped off a rock and the splash from my boot hitting the water made the trespasser bolt upright; and when he turned to face my revolver, I was staring at a giant apelike monster! The super huge Sasquatch and I stood staring into each other's eyes for what to me seemed life and death; it was decision time!

I relaxed my finger on the trigger, and for what had to be a minute or more, we actually just stared at each other. Neither of us did more than blink our eyes a few times as we, neither one, seemed to have an exit move.

I knew I was face to face with Oregon's biggest unsolved mystery, and the strange part to me was that I wasn't afraid. It was almost like looking at any other stranger, as I felt no danger; and the huge animal turned suddenly, and without even a cautious glance back, it proceeded to walk directly across to the other bank, the river only at its knees; and once there, it climbed hand over hand from tree to tree until it reached the top, about 10 feet above the river and then casually, right before it disappeared into the trees, it turned and looked down at me; then turned its back and bent down and "mooned" me. Then it straightened back up and waved its huge arm like it was swatting at an annoying fly. Kind of a wave of disgust, and then it walked into the trees.

As I walked up to the shack, I guess the creature had been really perturbed that I came up on it without warning, and it must have made sure it would have the last word, because a very large rock came from up above and landed against the side of my pail that held all my tools; scattering them everywhere! As I picked up what remained of my kit, I looked up once again and caught a glimpse of the Bigfoot's back; far way on a ridge.

When I told Ken about the incident he laughed and said that Sasquatch had scared him once and before it disappeared, he said he had "mooned" the animal, and apparently, it must have been returning the insult!

11 REVISITING SASQUATCH

Dean Winthrope Sr. ~ Riverside, California

The hiking season was almost over for anyone with any degree of common sense using the trails along the mighty Rogue River in Oregon!

Being as how I was on extended vacation and without a real schedule, I purposely saved my personal quest until well after the tourist season was ended, and I had taken several trips over the previous summer to keep myself in shape for the exertion that I anticipated making.

I had a motive for delaying this trip, not because I wanted it colder and wetter, which is not too hard to achieve in a state known for abundant rains. My reason for this excursion began over a year before when I last hiked this route, and it was an event during that trip that brought me back.

My home is in Riverside, California, where I spent over 20 years as a police officer, and I now found great pleasure in being by myself after so long being shoulder to shoulder with the masses. Also, since I was only 10 years from my retirement, I was scouting for a good place to move to later in life.

As I mentioned, I was here over a year before, and for concerns I have for the safety of the animals I found, and hope to find again; I will only say that I had left my pickup at a parking area past the lodge at Marial on the Rogue river at the road's end; which I had seen in last year's trip when I had parked close by the small ranger shack.

I shouldered my monstrously big pack and set out downriver to follow my memory toward my objective.

When I was here the previous summer, I had left Marial, and without disclosing the location of my discovery, suffice it to say, there is a place downriver about two or three hours where the Rogue River makes a sharp turn north, and it was a few minutes after that where I had twisted my ankle quite hard, and it hurt badly enough that I decided to take a rest. Searching along the river's edge for a good spot, I found an ideal and fairly private place to be comfortable.

Leaning back against a perfectly flat rock, and with my bare foot basking in the sun while resting on a cool stone, I was almost totally hidden from passersby on the trail above me; I was sure I could keep my ankle and foot from swelling.

The spot that I had found for my R & R was also underneath what was a high jut-out alongside the trail as it curved and headed down at a steep angle.

Here I was, overlooking a beautiful sand beach on both sides of the curve directly below me, and a beautiful wide beach covering a couple of hundred feet on the long, gentle sweep across the river. Out of coincidence, this spot was similar to the place about another mile downriver that was my goal for this trip. When I was last here, I had seen a Sasquatch just over a mile from this very spot!

There is an understanding among all Rogue River hikers that the Sasquatch lives throughout this entire region. Also accepted by all who enjoy this unbelievably beautiful country is that the animal of so much interest is protected to the max!

This I heard as a river guide answered an inquiry from an obvious "city type" who said something about shooting one; and there was little doubt about the answer he got back, which was a comment from the guide that, "No one would ever know, because there would be no evidence of the Sasquatch or the person who shot it." They would both disappear! So, my goal of photographing the Bigfoot seemed like it would be reasonable enough to me.

The swelling in my ankle was worse, so I must have twisted it more than it seemed, so now I began considering whether or not I could expect to possibly encounter the elusive animal in this area as well as the one a mile from here. I reasoned that just because I had seen one last year, that I couldn't expect that it would be in the same place now just awaiting my return. So the decision was made; I'd stay here if I could somehow find a way down to the river.

Getting my sock and boot back on and laced up again was painful, but I finally was ready for my descent. Behind the rock I had been resting on was a steep, but passable way down. I went carefully; with every step being made only when both hands were gripping the huge boulders that lay everywhere.

After an hour's effort I had descended over a hundred or so feet, and then I was fortunate to find a sort of trail that switched back downriver, and then in another hour I was standing on a large, sandy beach alongside a monstrous, deep, blue pool. The rapids were directly on the other side of this deep hole, and looking carefully into its depths, there seemed to be some sort of vortex that kept swirling the sands up and out of the pool.

Now, as I slowly limped along this stretch of sand, I was quite surprised that the trail above was not visible at all. I knew that there was a fairly steady flow of traffic; however, the typical hiker eventually gets into a sort of a lull where they settle into the steady hiking pace that it seems as though all focus is strictly on the trail; there could be dinosaurs in these mountains that they would miss unless eaten by one! In order to be able to see an occasional hiker on the trail above, it was necessary that I stand directly alongside the rushing river which was at its narrowest this time of year.

The area I was in now seemed a perfect spot to make camp since my ankle was swelling badly enough that I couldn't go further. The weather couldn't have been more perfect, and there I was; leaning against a huge rock in my bare feet, legs outstretched toward the river, and there before me was a beautiful canyon which was the source of a rather large stream that joined the Rogue with hardly a noticeable ripple.

As I lazily observed the water that appeared to head directly toward my feet, I wondered at the wide valley from which it came. I could see a colorful canyon that had it not been in such a wide opening, I may not have noticed its immensity. Perhaps this had once been formed by an ancient glacier or maybe a volcanic eruption placed mountains all around it.

I had fallen asleep, as all of a sudden I found myself reaching desperately for a handhold to save myself from falling off the cliff I had dreamed about. My eyes now open, my heart beating like a drum, and directly across the river a strange being was looking right at me! My sleep must have been very deep, as I felt no alarm and no fear; only the curiosity of waking in a strange place, and the "Where am I?" moment.

The creature staring at me could only be a Sasquatch, and I sat there without fear or panic, because I had seen one previously, so I knew exactly what it was! For some unknown reason, perhaps because I hadn't attempted to rise, the very large animal did not seem to be alarmed by my presence, so I didn't move.

The Sasquatch seemed large, but I observed the wind blew its fur in all directions. It was long, and must have been very lightweight from how it blew in swirls from the canyon breeze. The animal's eyes were yellowish brown and its paws seemed overly large with long fingers, and as it took a couple of steps, I could see a pair of huge feet. It had a nose so flat that it was almost like thick nostrils on a wrinkled, gray face, covered mostly with long, wispy hair. The animal's ears seemed smaller than what I would have suspected for such a large (seven feet ++) animal.

We were both staring into each other's eyes when I finally acted the way it seems we humans always do; incorrectly! Like a dummy, I let my excitement over such a tremendous discovery revert to a typical human trait; I spoke. The moment I did, I wished I could have been smart like an animal and that I could have had the patience to remain quiet. The Sasquatch reacted exactly as it should have; it ran like hell!

I spent that night wrapped in a space blanket, which when I retrieved it from my pack, it was necessary to first unpack that familiar object that excitement so often forgets; my camera! My return trip to my vehicle was spent in recalling every move in my trip and with every step, I winced in pain, which I felt I deserved by being so totally unprepared for the venture I went in search of in the first place!

12 WE WERE NOT ALONE

"Tank" and Tina Garner ~ Baker City, Oregon

My wife and I were hiking an area in Oregon's northeast in the Baker area where the hills at one time were covered with gold miners. This was quite a few years ago, before Baker's name was changed to Baker City. We had moved here 10 years before.

Doug from Portland, USA [CC BY 2.0
(https://creativecommons.org/licenses/by/2.0)]

We think we may have discovered what may be the basic of communication among the Bigfoot animals. We were hiking in our favorite hills, amidst a sometimes confusing series of trails that wound through these remote lands.

This particular day, we were in an area where there were a lot of trails cutting this way and that, and we had our first clue of the presence of Sasquatch when we came to a fork in the trail. We selected the path that turned to the right, and up toward a long deserted and crumbling mine building. Just after we took the trail, we heard a sharp "crack" like if someone hit a dead tree with a stick or a rock, but we just assumed it to be another hiker. Soon we came to our next split in the trail and we decided to explore the remains of an interesting looking, long abandoned, mine shaft. We made a left turn and began to descend the hill when there came a loud sound again!

This time we heard two distinct "clacks" similar to the first one. We sort of filed it as coincidence until we came back up the hill later and turned back the way we had come. Sure enough, there was the noise again, but only once this time.

Now we knew this was more than coincidence, so we decided to continue on our original course and we took the upper trail again. We were dropping through a fairly heavy growth of pines, and they were so closely spaced you couldn't see through them more often than once every hundred or so feet to the valley below.

We knew that we would soon see the very unusual spot where there was a large complex of buildings that took up the entire end of the large butte with three separate mines. The crumbling buildings were plastered with "Danger" and "Keep Out" signs, as the original owners still bore the responsibility for the safety of the general public, and their closure after the gold ran out had likely left the owners unable to handle the cost of properly closing up these shafts. A friend of our was severely injured when he broke through the floor in one of these mines, so we gave it due caution and steered clear of any more than quickly looking inside.

Besides, a lot of reports keep showing up in the paper and the online chat networks about the families of Bigfoot that are suspected of living throughout the massive area that is virtually littered with these abandoned mines.

It's really too bad, because back in the early 1900's or whenever it was that the price of gold no longer made it profitable enough for these miners to continue their operations; they just walked away, and most of them lost everything while holding on in hopes the market would recover. By the time gold became of value again, the existing tunnels and adits were no longer safe, and the tremendous start-up costs were prohibitive, so these super-rich mountains were now the home of the Bigfoot, coyote and bear; and yes, hikers!

After a bit more exploring and a quick lunch, we returned to our truck, and once again, every turn we made was to the quick accompaniment of a sharp knock or crack on wood. Keeping track, we deduced that there was a definite pattern; right turn was one rap, left turns was two raps, and even when we descended one very long trail that had six switchbacks in a long, steep descent, the raps proved to be accurate. Our concern was that the entire time we were out, something kept a loud account of every move we made, yet try as we did, the only times we saw our "lookouts" are three separate times were one of us would make a turn and the other watched from above.

Then toward the end of our descent, we caught a glimpse of a large, extremely dark Bigfoot down below us; waving up to what must have been another one that we couldn't see up above, because almost instantly the loud "crack" came from over our heads!

In a lot of ways, knowing that these massive numbers of abandoned mines can serve as homes to the Sasquatch is nice, since it seems as if this elusive creature is on its way to becoming our "state animal!"

Albeit, that it sure makes one nervous to know that your passage through these mountains is broadcast to all of this obviously large population of ape people. We hope they are vegetarians!

13 I ALMOST DIED AT DEAD MAN'S LAKE

F. Nolan ~ Missoula, Montana

I was standing at a counter for the Forestry Management in Montana awaiting information on a small wilderness lake shown clearly on my aerial photograph. Since my retirement, I was on a quest to fish the most wild and difficult to access pristine lakes in the state.

This particular lake had eluded me, as the only record I could find of its existence was by an old aerial photograph. Finally, I was ushered into a side office by an elderly man holding a sealed envelope. As we briefly discussed my quest, the mystery was disclosed; the man spread a map open and carefully peeled off a square piece of paper covering a particular area where my lake was clearly shown. There was another notation on the map itself, written in pencil that said, "Dead man's lake – Closed area."

Then, the man explained in low tones, reminiscing about the two hunters who had disappeared there while deer hunting years back. Numerous search parties, over several months, had been sent to the area, but the only signs of the men were a collapsible fishing pole and the two hunting rifles that were positively identified as theirs. The man said that the records noted that the hunters must have made camp by the lake and had likely enjoyed a meal of fish during their campout. A few items that must have been from their backpacks were found scattered, but all of the flyovers and search parties failed to find any other signs.

The old guy explained that the "area closed" sign was to protect the section and out of respect for the missing men. He said it was maybe time to again open the area since it had been 18 years and he discarded the attachments, so the map was again exposed.

I bought a copy of the map and returned to my motel. I had already packed my rig with all of my gear, and now I drove to the grocery store to stock up on enough food for a week or more. As I shopped, I pictured fresh fish frying in my pan.

The next morning I left well before dawn and after four hours of racing down the two-lane highway and another hour and a half on a two-lane dirt road, on which I passed only one farm, and never once met another vehicle; I felt right at home! In most of the remote areas like this, the people tend to live closer to the towns, as the snowplows covered the main routes first and seldom had responsibilities for the roads heading into the vast wilderness regions.

My heart was racing with the anticipation of catching huge trout that had never seen a human! With an inlet and outlet creek, this had to be a sleeper. I turned according to the direction I had been given and after swinging off the county road the words "Primitive 4x4 road" needed no further explanation. I began in 4x4 high, and for the last mile until the trail ended, I was going slowly in 4x4 low.

Judging how the rains and melting snows had almost totally obliterated the ruts in the ancient trail, the lack of human presence was obvious. There was a turnoff to my left in what may have once served as a parking lot when the place had regular visitors, and I parked there by carefully backing in close to the road out of habit from visiting so many popular fishing spots.

A quick, last minute check of my gear, and I hefted my backpack with everything from my collapsible fishing rig to my lightweight waders, and I strapped on my gun belt with 36 rounds of .357 magnum and checked my Ruger revolver to make certain it had six rounds; snapped the holster strap over the hammer, and hat on my head, boots laced, I was on my way.

The sun was up bright now, and the trail to the lake was still indicated by a sign that said "Lost Lake;" rusting with bullet holes that made its name almost unreadable. I wanted to make good time so I could string up my thin, one-man tent and make a semblance of a camp somewhere near the lake before dark, so I hiked at a rapid pace on the faint, grass covered trail. The grass was light, the ground hard, but the rocks were easy to see and avoid.

To keep up with the latest trend in television gimmicks, I was wearing my "tactical" sunglasses that I still wondered how they differed from my old ones, but maybe my "tactical" camp light would prove its worth. Just poking fun, but I couldn't find any "tactical" bear spray, so I made no effort to tread quietly as if I were hunting, but to the contrary, I didn't care if I kicked rocks on the trail. I occasionally broke twigs, threw stones at dead trees, and acted like a total "city slicker." No way did I wish to suddenly come upon a slumbering mountain bruin by being too stealthy!

Much later, I had made a rough climb up a long hill, and I felt that I should be close to my destination. Figuring that the lake should be in the valley that crossed below me, I felt my goal was in sight as I stopped to scan the mountains on both sides. The valley seemed to curve through the mountains, and I remembered what, on the map, had appeared to be a river that the lake I was seeking had a creek going through it and into that river. As I began my descent down the steep trail, I noticed movement on the trail across from me where it curved slightly to head along the creek.

It looked like a large, dark, brown bear, but I feared it might be a grizzly, which made it a whole other situation! I stopped and watched as the large animal was standing on its hind legs reaching for something in the pine tree in front of it. I assumed that it was after a beehive, because they love the honey; and then it pulled its paws back and turned to its right and walked down the hill into the valley.

As I continued down the hill, my mind was racing; something wasn't right! It took me several minutes before I could sort out what was troubling me, and then it hit me; that animal was on its hind legs at the tree, but when it left to go down the valley, it was still walking on its hind legs! Plus that, its arms were much longer than any bear I had ever seen, and I've seen a lot of them. Then I knew!

Now I've been all over the country, and this was the second time I had seen a Sasquatch! I should have known immediately what it was, as I have met a great many hunters, fishermen and hikers who have encountered them also.

Well, I must have walked faster, subconsciously hoping to see the Bigfoot again, when all of a sudden, I saw the shimmering of sunlight on the creek leading to and from this pocket on my left, and then I saw the lake! Maybe I should reference it as a large pond, but the darkness of its sixty foot diameter intrigued me. All around the beautiful pool was a wide fringe of rushes and tall swamp grasses of tan and light green, fuzz-topped plants. It looked like paradise!

With the sunlight no longer warming this narrow valley, a slight chill came through on the breeze, so I got to the chores of making camp. It didn't take me long to dig a fire pit close down to the water's edge, so I was not under any trees that a spark from my fire might catch on, and I was on the last hard graveled spot before the swampy area. I unrolled my very tiny tent and my waders; the rest of my equipment amounted to a small mess-kit and my tightly packed food supplies. I took off my canteen, but left my revolver on my hip.

Not anticipating any animal or human threats, but friends who have encountered Sasquatch, warmed that if surprised, they can be very dangerous. I always felt they told me that to keep me away from their favorite fishing holes.

Before dusk, I walked along the high ground bordering the marshy area to see if I could find the best path to reach the lake, but there was a thick area of marsh the entire length of the creek. The ground on the other side however, was rocky and full of trees right down to the grassy water's edge. How I wished I could be on that side, but I could not see any easy way, so my quest would be made in my hip boots, and I was glad I had them along!

The evening was usual for this type of camping spot; millions of mosquitoes! I ate quickly, and being exhausted, I was ready for sleep, so I squeezed into my extremely small tent and zipped it up quickly to keep from being in need of a blood transfusion by the painful little monsters. I really hate mosquitoes more than anything!

I awoke the next morning to a sound of "knocking." It sounded like a heavy stick being smacked against a hollow tree, and I thought at first it was a woodpecker, and as soon as it stopped, a similar rapping came from up the valley. I just lay there wide awake, trying to get my head around the sounds until I recalled friends telling me that this was the way the Sasquatch communicates.

Then I was up and moving quickly to get a fire going, as it was really cold! I had dressed warmly, but there is nothing colder than an early fall morning in the Montana mountains; unless maybe July in Minnesota; ha! I had a pleasant breakfast and I anticipated a fish dinner as I put on my creel and everything else I may need; including my revolver.

Having scanned the marsh enough now, I selected what I felt would be my best approach to the pond's edge. Being careful to not cast a shadow; I, with help from a fairly straight walking stick, carefully selected my path. If you've seen a typical marsh, this was it; grasses everywhere, but surprisingly, there were a few hummocks. You know; those thick patches of grasses that you can step on to walk across the softer areas. That is, if you're super agile and not carrying 30 extra pounds like me.

I did see a thin line of willow brush in an almost direct line to the pond's edge and it had what appeared to be one of those higher areas where it was normally quite solid. I could see what I assumed was a larger bush; you know, the kind that grew in the harder ground where it wasn't totally covered with water. The branches I could see were all dead, but they looked to be a couple of inches in diameter, and the bark had fallen away to leave their bright branches standing tall, and I could easily cast into the pond from there.

My walking staff was of little help in my journey, because it seemed as if I could push it down 'til it disappeared. As I followed the apparent ridge, there would be a momentary feeling of support and then my foot would sink. The muck would then close in on my ankles with a more firm grasp with every step, which caused me to take a more rapid path toward what I assumed to be high ground ahead where the large branches were.

My stick was useless now, but I kept it with the assumption that I could use it on the higher ground which I was rapidly approaching. Being as how my feet were getting harder to extricate with every step, I was happily approaching the first branch of the dead willows and hoping my next step would hit more solid footing.

I reached out and took ahold of the almost white branch and pulled on it to step up on a more solid hummock when the branch came loose in my hand. That was when I got the shock of my life; as the branch came to me I realized that it wasn't a branch at all. I was holding onto the antlers of a bull elk!

As the antler came to me, the rack suddenly rolled, and as it rotated, the other antler rose to the surface and I was staring into the eye sockets of the unfortunate animal. I instantly came to grips with the reality that this poor creature had met the same fate that was awaiting "yours truly!"

If you have ever had a near death experience, then you'll know what I felt like; I became overcome with fear! There I stood, but I knew that the muck had sucked my feet further downward in the seconds I had stood there in shock. I had lost all semblance of reality, and I was in a panic! I couldn't concentrate; my mind was flashing pictures of my funeral, the death of the poor elk, and worst of all, my last moments as I sank into the water!

I tossed aside these delusions, and forced myself to think; fast! I quickly pulled the elk rack toward me and broke one side off which came easily away. Next, I tried to use the heavy rack to pull at the grasses behind me, but it seemed as nothing growing in this bog had any root structure, as the grasses just came away. I had hoped to pull myself back, but that wasn't about to work, as they provided very little resistance; however, perhaps if I were able to float?

Then I had a flash that just in case there was anyone in this wild country that might be near enough to save my sinking ass, I had to try. I stuffed in my ear plugs and fired one shot after another in series of three with a couple second pauses; reloaded the cylinder and repeated the process until I was down to only two cartridges, which I kept. My ears were ringing as I strained to hopefully hear a human voice or an answering shot. I was fast running out of time and I had to try anything; I'll admit to being in a panic!

I knew my gun had to be dragging me down, so I quickly removed my gun belt, and making sure the hammer strap was tightly fastened, I looped the rig over the elk antlers and moved it atop the marsh grasses nearest me. I wanted my gun near where I could reach it, because my biggest fear of anything in life is drowning!

If I failed in my next idea, I wanted to be able to reach my gun rather than struggling for several frantic minutes paddling to keep my head above water and slowly drowning; I would end it quickly! Here I was struggling to survive, while wondering if I had the guts to really commit suicide.

So here I stood, sinking slowly into the muck that was steadily pulling on my ankles, and as I looked down, I had sunk deeper by inches. I recalled once mentioning to my friend Denny that I pictured this small lake as the "spot to die for." How that statement haunted me now!

Then the "little light went on" and I was frantically sawing at my left hip boot with my knife! I pushed the blade downward at an angle to avoid my flesh while holding the top of my boot. I fought to control my shaking hand, by the time I had reached the top of my ankle; my chin was in the water. Taking a deep breath, I plunged my head beneath the surface and I had to concentrate on cutting the narrowing ankle of the boot that I could not see in the murky water. I felt a sharp pain in my ankle, knowing I had flinched, and then I felt a slight movement as my left heel moved slightly. The boot was finally loose!

Working now with renewed hope, I made short work of my right boot, but with another painful slice. I could see a dark reddish tint in the water around me, but I had to act fast so I ignored the cuts as if it would help me any to know anyway.

I carefully rolled to my right side, extricating my left foot from its prison, and as my right foot began to experience terrible pain; then, it too came loose, and I quickly grabbed a handful of swamp grass with my right hand, gasping in fear as my face went under water! This momentary dunking caught me coughing, choking and gagging as I fought for air, and then I was going hand over hand; grabbing a handful of grasses, pulling until their thin root structure tore loose, while kicking frantically; holding my breath until suddenly my hands and knees were being abused by solid ground!

My fingernails were broken and bleeding, as well as my knees and toes; and some small cuts bleeding painfully on my ankles from my own knife, but I was alive!

As I sat on a large log in my camp, I tried to calm myself, and then as I was staring at what was almost my grave, I pictured that poor elk; as he must have experienced the agonizing death that I had faced only minutes before! I wondered how long he had been stuck there, and I surmised that he must have hit some solid bottom and just stood there and starved to death.

Then, feeling of sadness for a fellow creature hit me. I guess because I had been so emotionally involved with the whole experience, I packed up what I could do quickly, made sure I still had my keys, laced up my hiking boots, and without bothering to dress my wounds that were leaving a small trickle of blood in my boots, I headed back down the trail. I had a brief thought of trying to rescue my revolver by trying to hook the belt with a forked stick, but I heard my subconscious mind asking, "Are you nuts?"

So I looked at the pond one more time, and as I began to turn, something caught my eye, and there; staring back at me from where it sat casually against a tree was a small Sasquatch! How big it was, I couldn't tell, but it could only have stood as high as my chest. Then, a high-pitched screech like whistle echoed from somewhere, and the little guy took off like it was late for dinner!

I made record time back to my rig and I drove steadily until I got back to town. I bought all sorts of bandages and ointments at a drug store, with the clerk not asking any questions as I bled on the floor and a bit on the counter. Then I stopped at a motel for the night, called out for a food order and proceeded to bandage my physical wounds and lick my mental ones while I made a decision for the future. From now on, I'm fishing only from a boat!

14 BIGFOOT / LITTLE ME

Jim Lymans ~ Fortuna, California

Sasquatch was the last thing on my mind as I left Sacramento. I had sustained a severe injury at my aerospace job and the company was unusually generous with my check for severance and with my monthly award to add to my forced retirement, I had to decide where to put down the roots of retirement, so I thought I would spend a time relaxing, as the last six months I had spent in recovery; there were people everywhere, and now I wanted nothing more to do, but be alone with me!

I got really lucky with my search, as my broker had a place immediately available; a nice couple in Fortuna, California were leaving for a mission for their church and I rented the furnished home for a year. Two weeks later, I arrived with my Highlander packed to the roof with everything I'd need for months to come.

The local realty company representative was there when I arrived, and he was kind enough to also have a packet of local features; stores and what-not; the things that interested me most were the personal notes written for my benefit; places for the most solitude where I could get away from people.

After a week, I was all moved in and rarin' to get out and enjoy a nice, long hike. My backpack full and my destination selected, I soon found my first goal, and parked beneath a group of giant redwoods not too far from Humboldt.

Daderot [CC0]

Boy! I had a lot of paths on which to embark, and figuring as to most likely walk every one of them before I was through, I chose the one on the left; and in a couple of minutes I was in another world! Not that my hearing is all that great, having spent so long around jet engines, but very soon, the sounds of the Pacific Ocean blotted out all else.

That's when things changed; I caught a movement out of the corner of my eye and after a casual glance, my mind told me something was not right. I recalled again the figure I had just seen when I first looked, and I quickly turned back to look again, but it was no longer there.

I had obviously accepted what I saw, because I was expecting to see a human being, and it was only a split second later that my brain did a recall; what I saw was not human! A humanlike shape; yes, but not a human.

Not even thinking about being afraid, at least not consciously; I found myself rummaging through my pack as I was quickly walking toward where I had seen the figure, and soon I found myself carrying my large "push knife." Probably illegal in California, but what isn't? Anyway, I had owned it since Desert Storm, and I always carried it out of sight.

I hadn't slowed my pace, and up ahead I could see a grassy area on both sides of the path, and hopefully my tracking skills were still intact. As I came up on the area in question, I slowed to search the dip in the path, and sure enough, there was a set of footprints, and they had heels plainly visible.

Then I noticed that one of the boot prints was half obscured by another smooth print. There appeared to be a large heel, but from something, or a very large someone with bare feet! Looking more closely, I could also see toe prints that obliterated parts of the boot print entirely.

Now my thoughts flashed back to what I had seen, and I knew for certain that the creature was not human! As my mind began correcting its wandering from trying to see what was natural to my memory and to replace the "automatic human association" with reality; I knew it was a Sasquatch! The fuzziness was gone now, and the strangest sensation hit me. Now that my brain was through trying to rationalize something that it had never seen before, the first vision I had of the animal was now more plainly pictured in memory, and I picked up my pace to overtake the creature.

I was walking as fast as I could, as I dared not give away my pursuit with the noise made by running. Besides that, I had a pretty good idea that this thing was much larger than I. The term Bigfoot is kind of a misnomer, as I think the picture my mind had would maybe consider changing its name to "Bigfoot with damn big body!"

I stopped a couple of seconds at another semi-damp area on the trail and there was another print; very plainly showing a large big toe and four others, but the big toe was not all that much larger than the other four. The clarity of the print and the way the animal placed its foot down didn't seem to this old deer hunter that it was in a hurry; so it was likely unaware that little ol' me was in hot pursuit. Maybe if it had, it may have waited to make an introduction and "killed my ass!"

I just could not have been rational, as I was going as fast as I could now, and all of a sudden, up ahead I could plainly see this very large, brown creature with its beautifully shining coat as the sun found a way through the Redwoods to light him up like a spotlight. Yes, I say "him," because he was not wearing clothes!

About that time I had closed the distance between us to where his golden brown eyes could be clearly seen glowing at me, and I had the thought that, "I was too damn close!" That thought quickly disappeared though, because that giant animal suddenly bolted into the bushes beside the trail and into the forest without my ever seeing it again. Soon the "thumping" of his footsteps was gone as well.

On my way back to the house, I stopped into the local camera shop and bought a nice compact camera. As the clerk was completing the registration, I mentioned that I had gone hiking today in the giant Redwoods and had the strangest experience; the man stopped me and said, "I'll guess you saw Sasquatch!"

Anyway, I'll send you a photo next time!

15 SASQUATCH AND THE LOST GOLD MINE

John T. ~ Grants Pass, OR

Publishers note: This submission was sent to us by a man whom we had last heard from over four years ago. He had been searching for a gold mine that his great grandfather had found very near the Almeda gold mine on southern Oregon's famous Rogue River. The Almeda was a huge mining operation.

He had purchased our book "Hiking Sasquatch Country" in which we had referred to a newspaper and magazine articles where the author Zane Grey had brought to light, the incident about the murders of four men from the Almeda mine who had cut out on their own in search of a gold strike. Zane Grey reported from researching historic records, that the four miners had been "torn to pieces" by some unknown monster! The bodies of two men were recovered along with signs and articles from the other two among the "giant" footprints that covered the entire area!

This man's great grandfather had reportedly been employed at the Almeda mine during this incident. From the way it was told to us, in the confusion that surrounded this long search, things at the Almeda were in a turmoil, and our submitter's grandfather and a couple of his buddies slipped away for a while and they struck out on their own and supposedly found a small strike on the other side of the Almeda mine's claimed ground around the mountainside.

The man's name (let's use the name John Thorne, which has a special meaning for the family) had sent us this story after our hiking book helped him to finally find a way by vehicle to get to the upper mountain that led down to the Almeda.

We know how hard it was, because it took us three previous attempts while we were investigating this story! The only other option was to float downriver, but the mine is private property, so a "back way" was desirable for our purposes and for John's. There's nothing more suspicious than a tied up or beached raft on a privately owned gold mine property.

John began by calling us with his thanks, but then, it had taken a couple of years before he could relocate his great grandfather's story, because his grandfather had since passed away, and it was he who had inherited the records. Had we not since moved from the area, we would like to have accepted his and his wife Millie's kind offer to accompany them on this search; it would have been memorable!

The area begins miles above the raging Rogue River, and with the use of old logging roads and the main Almeda road, the road is blocked off miles above, but it makes for a great hike down.

So here is John's story:

I have waited many years to tell of this event, and I have often sat down to put it in writing, thinking no one would possibly believe me, so I always quit. Then, after I broke my leg a short time ago, part of my reading material my wife collected for my recovery included a couple of your "Sasquatch" books. After reading two and insisting my wife order the rest, I was amazed at how many others had similar experiences to mine with our Bigfoot animals.

According to great grandfather Hiram's payroll sheets, he was one of the foremen employed at the mine, and from some rather sketchy notes, evidently when the four men left, it created quite a lot of concern at the Almeda, because the management correctly reasoned that if the four of them reported finding gold on their own, the Almeda could easily have lost their entire crew of three of four hundred miners, and with the Rogue River, from there to the Pacific Ocean having only a few claims; there would be a mass exodus of available replacements throughout the nearby towns as well.

Grandfather's diary had a crude map sketched on a separate piece of paper and inserted in the book which he carefully traced onto a sheet he could carry with him.

Millie and I set out on our search with our overly large (according to Millie) backpacks and our generous-sized monster dog named Sarge. He looks a bit scary but he wouldn't harm anyone!

So off we went with our revolvers strapped on, and your book took us exactly to the upper adit "C". Just for fun, we took the extra couple of hours to hike all the way down to "A" level and scared up some wild turkeys and a few deer; and even a slinky coyote. As your book stated, everything was locked up, so we could only peer through the steel gates covering the three adits and wonder.

The buildings, of course, are long gone, but we checked out some of the massive generators and such, and then beat it back up top where we had hidden our packs; and it was then that I pulled out Grandfather's sketch, and we were amazed at how, after all these many years, the several landmarks he noted were so easy to still identify.

The trees of course, were much larger, but the trail through the large boulders was still easy to find, and with the trail obviously the best way for animals to get up and down through this tumble of rocks was plainly the only way. The meticulous directions that Great Grandad had taken such time with had obviously been rewritten from his field notes, as at some points it got hard to interpret.

We were almost stymied by one turning point, and we stood there for quite some time trying to decipher the landmarks, when finally we spotted the stone pyramid he had referenced;

we finally realized that lightening must have dropped an absolutely tremendous-sized tree directly on top of the rock, almost obscuring it beyond recognition. Had Sarge not split to chase a squirrel, we would likely have passed it by. Anyway, ducking under and squat-stepping about 20 feet, we emerged directly facing the trail he had said was easily missed.

We walked the trail along a fairly wide ledge that wrapped around this cliff, almost as if it had been created by man, but it was simply a "gift from nature" across this massive boulder field. Looking up, which we often did, the entire slope above us for over 200 feet was nothing but rocks. Where the trees began again, there were many rotting remnants of the majestic evergreens that had ridden down the tremendously large landslide.

We soon came to the ridge indicated on Grandad's map and found ourselves walking beneath a wall of rock that was about 50 feet above us, and although there was a trail, it was very narrow; and with Sarge first, then Millie, I was bringing up the rear when everybody stopped short!

I immediately saw what had caused the halt; there stretched on the slope below us was a large, dark shadow that looked like a human! I looked up into the bright sun, and shielding my eyes with my palm, I saw an ape! I heard Millie gasp and choke out something, but I couldn't understand as I was trying to step down off the trail to an outstretched sort of shelf made by two huge rock slabs when I slipped, and down I went; right on my rear!

As I was sliding head first on my back, I remember the image of a large, hairy apelike animal, but I don't remember it being slouched over like the gorillas you see at a zoo.

Fortunately for me, I landed in one piece, still upside down on my back, but I could still see what I finally knew to be a Sasquatch! This was the first one I had ever seen, and I noticed Millie still trying to shield her eyes to view it better.

She told me afterward that she had been too worried that I might slide all the way to the bottom, so she lost track of the Sasquatch for a bit, but when I landed safely, she looked up again and the big ape was still watching me also. Then, as I struggled to rise, it was gone! There seemed to be an entire plateau up there, but we could never see a way up as much as we searched for one.

I had a lot of scrapes and bruises, but nothing broken. That is, unless you count my wallet; my revolver had come unsnapped and it was nowhere in sight. It would have been impossible to find it in that massive jumble of boulders and broken rocks strewn out below us, so I had to forget it. As the three of us got going once more, it took a bit of coaxing and dragging with Sarge, 'cause he wanted to go home!

We were trying to hurry now as the sun was low on the horizon, and since we were high up on the mountain, we were in sunlight still, but some of the valleys below us in the distance were already darkening.

We had tried to allow ample time, even with our excursion down to the lower mine adit, but now we began to hustle. Then Millie, who was our map interpreter called out to "look!" There directly ahead, and where the slide ended, the large forest of monster pines stood as Grandad had noted as a key reference point.

Following the detailed instructions at this point, we proceeded to the center of this island of 100 to 150 feet tall monuments of such incredible beauty as to make one stare in awe! We were both impressed, and Sarge made a courtesy anointment of the nearest pine.

In 15 minutes, we had dug a fire pit, lined it with rocks, cleared the area of pine needles and gathered enough downed pieces of timber to have a small, but wonderfully attractive campsite. Making certain there were no trees directly overhead, and having dragged several large pieces of wood over for backrests, we were sitting on the ground roasting weenies; some quite squished from my landing on my back, but all tasted great. Soon, Sarge was stretched out fully, having eaten all he could, plus a bit more!

This was our destination that we had both been concerned about by great Grandfather's memory and his accuracy. Since I had never had the pleasure of meeting him, I wondered aloud if maybe perhaps he was looking down on us now; and at that precise moment when I mentioned it to Mille, a large pine cone fell near our fire pit and bounced alongside! How unnerving, but I took it to be a "sign" and I still have that pinecone in a special place in our living room.

Well, that entire evening was the most enjoyable time one could ever possibly imagine! Sleeping on the ground in a light sleeping bag, while quite comfortable, was something some of us can think back on, but to do so when one is on in years can be a totally different adventure. Although it wasn't something I would jump to do again, the act of getting up in the moonlight to stoke the fire, with the light wind carrying the distant calls of far off coyotes, was an adventure I shall never forget!

The mountain air was rudely cold at five o'clock in the morning, but the sun gets up early to light the mountain tops, and the welcome fire soon had these old bones heated enough to move once more, and after a quick bite to eat, Millie and I were once again poring over the old sketch of where the mine was supposed to be. We had no idea whatsoever of what to expect, but according to the map, this was where great Grandfather and his two friends had made their main camp, and their gold mine had to be within yards of where we now stood.

Millie and I carefully circled the smallish clearing, looking for what the notation said was a small slot behind the tree, which we were diligently searching for, and we were right along the edge of the three foot high ridge of rocks that arose gradually to a ridge above us, and then the cliff shot straight up several hundred feet.

It just had to be here; and we were both now beginning to have our doubts when a large marmot of some kind ran almost over Millie's foot; she shrieked, I jumped, and Sarge charged like Custer, and they ran straight for a tree neither one could climb; and just as I thought they would collide with that tree, the marmot darted right through the bush growing alongside, and both animal and dog disappeared!

Millie and I were there in a heartbeat; what had appeared to be tree trunk all the way across was an illusion; the tree had grown out around a protruding rock by the cliff and left a good sized opening before it straightened itself out once again, but in the interim, a bush had grown into the cavity, and together, they effectively concealed a narrow slot that went directly back into the rocky cliff. The tree had obviously grown a lot bigger over all of these years, and realizing we had almost passed it by, I was visibly shaking with excitement as I stepped up and around the bush; Millie was right behind me as I carefully and apprehensively proceeded further into the slot; with the walls on both sides gradually rising higher and higher until we now could no longer see anything but cliffs and treetops.

Then, the slot dead-ended at a gold mine! There on our right was a large hole into the surprisingly fragile cliff. On closer inspection, it appeared that the entire cliff was made of some sort of fragile rock that broke away easily as I hit at it with the tempered steel rock hammer's pointed end. I had purchased it for this trip as a "just in case" tool.

We set our packs down and retrieved our lanterns, which together, totally illuminated the crudely made gold mine. We were both trembling with excitement and even Sarge was dashing to all parts of the cave, smelling and scratching at signs of whatever the current resident animals were.

We were so very excited, that even to this day, I can vividly recall the trembling I experienced at really having found what I so very much wanted to, and it was one of those, "One in a million" moments!

The day was still young when we were carrying a supply of firewood, and the fire pit that my own great grandfather, whom I had never known, had cooked in and slept here and dug into these very walls. The realization was overwhelming to me, and the feeling of exuberance was like nothing else!

So there, long into the night by lantern light, Millie and I chipped and pried into the small lines of gold veins, that while not large enough to get any big chunks of gold; there it was piling up in the larger pocket of my backpack to be ground out later on in the comfort of our garage.

By the time we had set for our return, we both found it hard to leave, even though a return trip would be super-easy, we were relishing every moment of our time here!

When we got back to home to "boring civilization" the following day, we called my dad, and he and Mom came over to see what we were so excited about; they too were almost overwhelmed with our discovery!

Dad wanted to come along, but he's confined to a wheelchair, but when I brought that to his attention, he seemed to have totally forgotten, and his shocking answer was, "Okay, I'll crawl!"

Well, this adventure was over three years ago, and our entire bubble burst two weeks after we arrived home, as I fell off my pickup bumper when unloading some bags of lawn seed and I broke my leg in two places. My healing has been slow, as it's been three years now, and we hope that maybe by next year I'll be able to make it back up to that memorable place once more!

We were discouraged against filing a gold claim on the mine, as my dad had mentioned hearing Grandma say something once about this find; that great Grandfather had been hesitant to file a claim back them out of fear that the Almeda may contest the claim as it is just over the mountaintop from their mine. My father felt much the same, even though the Almeda has been closed for so many years. It would be a shame to have my claim contested, because it would not only be so crushing, but I couldn't even go on the property again. So we decided to let it lay as it is, and if we should strike it rich, the gold will still be fun to spend, and being the owner of an honest to goodness gold mine will have to be sacrificed!

16 TRIBAL SASQUATCH

Del Barton ~ Skamania County, Washington

I was scouting for an easier way to get into the place I have hunted deer for the last two years, and it was about a month and a half before the season, so the weather was beautiful!

The reason I was in there way ahead of the spring opening, was because our camp area was hit by an early snowstorm last year, and since weather in the state of Washington is quite unpredictable around deer season, I was looking for an easier route.

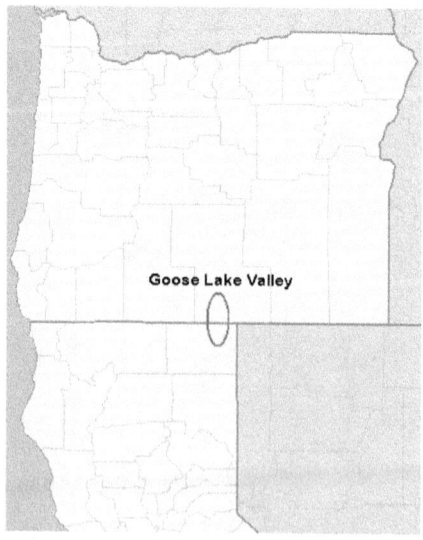

Orygun [CC BY-SA 3.0 (https://creativecommons.org/licenses/by-sa/3.0)]

I was in Skamania County, which is, for those unfamiliar with my state; just north of the Columbia River which separates Washington and Oregon all the way to the Pacific Ocean.

I had my dog, Jasper, along, and I parked where I knew there was an old Indian trail leading toward the Indian Heaven Wilderness area.

Bear Lake in the Indian Heaven Wilderness Area

I crossed on a direction sign indicating a trail to the Trapper Creek Wilderness area, where a friend of mine had gotten lost during last year's early blizzard, and he must have broke his leg or something, 'cause they didn't find him 'til the next spring, and there wasn't much left after the coyotes had been feasting.

Jasper was happy to be along, 'cause dogs can't go during hunting season, and he'd go off, first one way, then the other, except after we got in where the trees grew taller and the brush thicker, then he stuck pretty close by me!

As we followed the well-defined trail, it contained a whole lot of tracks, and in the softer areas, such as where the trail dipped, and there were traces of water on the sides; there'd be lots of different animal sign. Squirrels, weasels, lots of deer tracks; some elk, and then I found something really different!

"Indian Heaven Wilderness Area" ~ English Wikipedia [Public domain]

First I thought it was maybe a bear track, because all I saw was the back of the print with the front being in the grass. This trail had been used for centuries, so I learned, where I went to school in The Dalles, Oregon, there were a lot of old timers who talked about the creatures at "Ape Caves," that were supposedly like the Sasquatch that we have today, only they were said to be smaller according to the Indians.

Note: "Ape Caves" is notated on all area maps and there are a lot of weird stories about that I purposely avoided, because I didn't really want to know! It's scary enough in these dark woods that I don't need to add to it.

There are still lots of Indians around, and also a lot of Sasquatch too! I've seen them every other year or so during hunting season. Then after a few days or so, you don't see them anymore. They probably go where we can't hunt; animals seem to know.

My hunting pards think it's because they go off in some of the old lava caves and hibernate, but by then, the snow is so deep up around those mountain trails you can't get in there to look for them anyway. When I saw the first Sasquatch footprint a while later, I wasn't too shook up, but it's always a bit scary when you come on a footprint that's a whole lot bigger than your own boot print! That's what I was looking at, and the prints had cut in from the woods on my left and were going straight along were we were heading. I wasn't worried, because I always carry my same gun as I did in Vietnam; my old Colt '45 pistol. I can't shoot for shit, but those critters aren't all that much bigger than a man, so I wasn't scared so much as just damn nervous!

Even though I'd seen these Sasquatch a few times, I'd never been so close on a trail to one before. All of my sightings were quick and far away, so I knew I wasn't too far behind now, because I stopped to take a closer look at one of the more plain footprints, and as I bent over looking at one particular track, it was slowly beginning to fill with water. That indicated what I really didn't care to know; it was close up ahead of us, and Jasper was really getting nervous now!

When we're out like this, he always stays close by me and the only time he'll take off running is after the occasional squirrel, but he stops when I whistle. Now, he kept so close, I kept bumping into him, so he was obviously upset, which was also plain to hear with his low growling.

I was beginning to be even more concerned, because those footprints we were following looked about one and half times bigger than my boot print. Normally I'd have turned back already, but I had to be only a mile away from the cutoff trail that I had come here to mark with flags for our upcoming hunt.

We had first missed the trail during last year's hunt, because a light snow had come in the night before and was just enough to make us miss it entirely. I planned to tie several pieces of orange plastic flagging along the trail at eyeball height, so this season, we can save over two hours' time, and that counts a lot when the sun's setting earlier!

By now, I knew that we had to be too close to give up and turn back. I heard a distant series of barks, and it was then I realized that Jasper was gone! My immediate fear was that he may have run into the Sasquatch, so I hurried down the trail.

Now I was pushing through the dense underbrush as I made good time toward what now was a mixture of playful, short woofs. Thinking Jasper may have run up on another hunter, I went faster.

Pushing through a large bush, I came out in a small clearing about a hundred feet across, and there was Jasper, excitedly yipping and whining at another strange animal that I couldn't wrap my mind around. It resembled the Koalas I had watched on "Stuarts Outdoor Adventure" series, but it had more of a look of a small, fur covered human. I had never before seen a young Sasquatch, but I knew it had to be one! I stood there, having snapped Jasper's leash on, in case he decided to act on his own.

I had been too absorbed to think beyond the present, but Mamma Sasquatch must have decided it was time to call a time out! The entire forest was suddenly booming loudly, from the sound of what must have been a large log or branch pounding against a hollow tree in the blackened and dead remnants of last year's forest fire.

Thinking back to the reaction of the little Sasquatch; I wish my own kids would have minded me as well when they were young, because it took off like a speed skater; straight for a large boulder by an area of dark green pine trees.

I was staring intently, wanting to mentally record that experience, and I was rewarded by a very large and very hairy (If I'd been any closer, I'd be saying scary) Sasquatch. Jasper suddenly emitted a low growl and went behind my leg.

I couldn't' tell if it was male or female, because I was trying to mentally mark the place on the tree where the animal's head came to. Afterwards, once I convinced Jasper to come with me, I was able to determine the animal to have been over nine feet tall! I knew a lot of fellow hunters who had often seen these animals in this vast wilderness, but this must have been one of the biggest! I figured it must have been the young one's mother, but being that big, it may have been Dad!

Well, I marked my trail as originally planned, and later, during hunting season my partners and I were glad to have the flags, because we arrived during another freak snowstorm.

When I write this, I wish I could give more of a description of the Sasquatch, because it was so similar to flat-nosed brown bears, only taller and thinner. The fur seemed to be dark brown and tight on its back, while its belly fur was light tan and gray, and long and wispy; even under the arms and over the ears.

During our four day hunt, we were fortunate enough to tag two very large bucks, and as we were dragging them out, we crossed three sets of fresh tracks that could only have been made by Sasquatch; one very large, maybe like a size 16, one maybe a size 10 or 12, and a smaller set, that made us all laugh, as its tracks went first one way, then another, and it seemed like Sasquatch parents must have the same problems as humans in raising their children.

17 HOW DO YOU SPELL FEAR? ROGUE!

Brian and Samantha ~ Merlin, Oregon

When the huge power plant facility in San Diego that I had been working steadily on was finally completed, Sam and I decided to take a break. My older brother and his wife invited us to stay with them in Oregon, where they had moved to 15 years before. I knew that they would attempt to "sell" us on the idea of moving to Oregon, but we ruled that out because of the rain. We did look forward to an extended visit however, as it had been six years since we last saw each other.

We arranged for neighbors to babysit the house and headed for what turned out to be an adventure none of us will ever forget!

Jerry and Annie have a very nice home in the town of Merlin, Oregon, and they live, as Jerry says, "A stone's throw away from the famous Rogue River!" This was our first visit to their beautiful home; although they had been down to visit us several times.

Jerry and Annie had raised two great kids, but they lived elsewhere, so it was just the four of us in their huge home. They lived off a place called "Crow Road," and I could see why. Every day, we awoke to several of the large birds raising racket in the woods behind the house; and when the noisy critters weren't calling, we could hear the pleasant sounds of the water trickling over the rocky creek bed.

By the fourth day, both Sam and I were actually seeing ourselves retiring to this country – however, Jerry had been planning a surprise, which in retrospect, seemed designed to sell us on this area.

Anyway, there we were, following Jerry to the large detached garage where we were introduced to a complete assembly of river rafting equipment, and then Jerry and Ann told us that we had reservations to go rafting in three days.

While I knew that the rafting was a hobby of theirs from the photos, letters and phone calls over the years, we had hoped without asking, to be able to experience the adventure of rafting the world renowned Rogue River! We already knew that people from all across America and many from other countries came here for the rafting experience, as this was evidently one of the top four or five most famous rivers in the world! That was why Jerry had to make reservations so far in advance. You couldn't just stop anywhere and launch a raft.

So, we went through preparations, and both Jerry and Ann covered all of the safety precautions we had to legally know, and then we spent a few hours discussing river etiquette, and then a lot of time discussing the ride. How to paddle, bounce, cling to the raft to save your life, and minor items like that to make us relax and look forward to coming out alive; now I began to get scared!

The second day before our launch, I drove our rig following Jerry on a several hour drive on the north side of the Rogue River which I couldn't even see because we were up on a mountain road. Several hours after we started, we arrived at the town of Marial, which consisted of a lodge and several other buildings, including a government building used by the forestry department as an office away from their headquarters.

Inside, we had lunch as Jerry confirmed our room reservations for later, and then we found out the need to bring his pickup here. We were to launch the raft near Jerry and Annie's home, float down here and stay overnight, and load up the raft for the return trip.

It all made sense then, because it was still a long distance to the Pacific Ocean, and although Jerry and Ann had made that trip several times, they figured to "break us in gently."

Even though it was a long round trip that day on a pretty rough dirt road, it was really an adventure that we city slickers will always cherish; what a real taste of a totally wild river!

Time passed quickly, and it was soon time to go. We launched early in the morning according to our prearranged order.

Where many rafters put in at places Jerry pointed out, like Grave Creek boat launch, we had launched well before it, and Annie pointed out the old Almeda gold mine that once had round the clock shifts of gold miners. Hundreds of them were employed on that mountain, and how beautiful a location it was.

Our first destination was the Black Bar Lodge, located appropriately on its' namesake, "Black Bar!" The bar was underneath a very turbulent area, but Jerry skillfully maneuvered the raft to shore and we spent a very enjoyable evening in our rooms listening to the rushing river passing by. At dinner in the pleasant lounge, one of the patrons regaled some of the diners with tales of life and death on the Rogue River. This included the mysterious murder of Mr. Black who was found shot to death!

We hit the water at our leisure the next morning after a short walk around the premises to get the stiffness out of our bodies. As we made our way, Jerry was constantly giving instructions on how to lean, paddle, pull against the current, spin around to avoid rocks and obstacles, and it seemed at times to be a constant battle for survival! At other times it was so relaxing, one could easily fall asleep; that is if the possibility of drowning wasn't uppermost in the mind of yours truly!

With the speed of the river determining the length of time one could enjoy the scenery, we made many stops to get out, stretch, and since we were not in for the full trip that some of these hundreds of rafters were, we could take the extra time to stop and enjoy walking along the beautiful shore on both sides of the very wide river.

We all preferred the south side, which is where the Black Bar Lodge is located, because there, the shoreline is much less visited. On the other side is a hiking trail that courses the entire trip all the way from the Grave Creek boat launch to the Pacific Ocean in Gold Beach! The road we had traveled to Marial ended just yards from the Lodge, but the foot trail continued along the shore.

We had briefly hiked on the shore at Rainie Falls and dug around in the shallower pools looking for gold.

The falls were named for early pioneer, R.J. Rainie, who lived in a cabin there and panned gold and supplied fresh fish to the area. He was found murdered, and it was assumed it was for his gold cache!

We had made camp on the south shore, not too much further down, and we stopped earlier to take some time to explore a particularly scenic area beneath some high cliffs and among a small group of very large pine trees. Because of the severe fire restrictions, Jerry chose this location because it held one of the very few remaining "legal fire pits." It was solid-poured concrete and sat alone near the shore.

We camped early enough for us to take the time to hike back through the thick forest of pines. These magnificent dark green giants stretched so high that when we stood in the wide area in the center of the group, it seemed almost as if they had once been planted around a cabin, but we ruled that theory out, because that was too far out to be realistic. Annie explained that the idea was certainly romantic, but these particular trees were likely here and growing around the time that Amerigo Vespucci first named America.

As we exited this thick forest of giants, we were facing the cliffs that were much nearer to us. A group of bats were exiting their dark slots in the towering cliffs that blotted out the setting sun so thoroughly that it was pitch black in many of the narrow canyons that stretched along the river.

It was then that as Sam and I were holding hands, for better stability against the constant wind gusts when she suddenly jerked my hand so hard I almost lost my footing. I was about to ask why, when the answer appeared directly in front of me, about 100 feet away.

There, staring back at me was a very tall (about seven feet high) creature resembling a painting I had seen at an outdoor tourist souvenir tent a few days before! That painting was the artists' conception of a Sasquatch. I remembered it right off, because I had contemplated buying it, but my better half discouraged me. I will always remember the eyes on this creature, as we stood staring at each other; it seemed like long minutes, but amazingly, it could only have been seconds before this large, brown hair covered humanoid shape disappeared into the dark brush behind it; there really are Sasquatch!

By this time, Jerry and Ann had been alerted by our total silence and lack of movement, so they came around to our side of the trees barely in time to see our visitor as it reached up and grabbed a large limb of the nearest tree and up it went! We could see only an occasional glimpse of the brown object heading upward, hand over hand; the only sign of its passing was a light sprinkling of pine needles and small pieces of bark falling like a dry rain on us; several of which landed on my glasses and in my mouth. For that reason, I was convinced what I had seen was real; I could taste the fact!

The four of us instinctively retreated to the open ground behind us in order to have a view of the upper tree, and no sooner had I been able to again catch a glimpse of the creature, it launched itself effortlessly to the next tree, and then the next, and finally it swung over to the cliff and was gone! Afterward while sitting around our small campfire we repeatedly went over the details of our most exciting trip ever!

The next night we made camp at Winkle Bar; where the famous author Zane Grey had a runway for his single engine airplane. He owned a cabin there where he often fished and spent time alone relaxing. He even wrote the novel "Rogue River Feud" one summer while there. We purchased a copy of the book after we got home. We discussed that fact, and all of us asked aloud at about the same moment, "I wonder if Zane Grey saw Sasquatch?" I'll bet Sasquatch saw Zane Grey for certain!

After another day on the Rogue and a night at the lodge at Marial gave us a ready audience to discuss our Bigfoot experience. The proprietors had heard hundreds of stories of sightings and encounters, but they were all ears, because of where we had met this creature was a first ever sighting in that particular area, at least from their records.

We are currently making arrangements to put our house in San Diego on the market, because that experience was more excitement than either of us ever had, so we came to the conclusion that we would rather dodge bouncing rafts full of tourists and mountain monsters than the California freeways!

18 LAKE OF THE WOODS BIGFOOT

Darrin W. ~ Salt Lake City, Utah

Far below the summit of the mountain on a steep hill, a massive elk lifts its tremendously large rack as its eyes home in on a familiar and accepted forest dweller, but due to its resemblance to humans, suspicious and usually avoided; Sasquatch!

The eight foot tall creature passes by with hardly more than a glance at the fifteen hundred pound, statuesque sentry of the forest, and ambles its way down the narrow gully toward the wide pond below.

The giant walks stooped over, more like a gorilla in a zoo, but it does not use its knuckles as assists in its travel. The long arms occasionally flex outward, but seemingly more for balance, only occasionally reaching out to a bush or tree for an assist.

The winter has been long and harsh, and a meal of marsh grass and lilies is on today's menu. Perhaps a handful of the succulent water chestnuts will add to its pleasurable repast.

Far above this scene, a bald eagle soars soundlessly over it all, and its shadow passes over a crouching mountain lion as its eyes are affixed on the newly arrived mule deer fawn as she grazes on the new grass.

Into this serene wilderness, thanks to modern transportation, enters man; noisy, constantly chattering and disruptive to the peaceful atmosphere of one of the last truly pristine areas in the state of Oregon.

This particular couple has recently arrived at the "Lake of the Woods" campground where they had long before made reservations.

Introducing to this scene; Darrin and Danice W. We'll let Darrin take it from here, as it is their story.

I am taking an extended leave from my career as a long haul trucker. I have been to all 48 contiguous states in the union. Now that I have some time off, Danice and I are exploring places that we both traveled past, as the company allowed for her to ride along on many of my long haul trips.

We had admired Oregon's beauty from a distance, and now we pulled into the space we had long in advance reserved for the summer at Lake of the Woods campground.

United States Forest Service [Public domain]

The camp hosts, Roger and Linda, made us feel welcome, and after we parked our fifth wheel and trailer, they helped us make our area into a comfortable space like our own backyard. After a tour of the beautifully maintained campground and leisurely relaxing in our outdoor lounge, we slept like never before!

After breakfast the next morning, we packed a lunch and canteens and set out to check out our new summer home. We were wearing our brand new, but terribly stiff, hiking boots (lesson learned) and we followed the well-traveled lakeside trail.

A myriad of paths crisscrossed throughout the areas surrounding the lake, and we picked one that angled very slightly upward, and we could see above us that it appeared to switch back and forth as it seesawed up the lightly forested slope. The going was easy and few people were out from what we could see from our higher vantage point when we met another couple coming down.

Before we could even get out a hello, the man blurted out, "We saw a Bigfoot!" I met Danice's gaze, and I could tell she also figured it to be a gag. The lady however, joined in, and after introductions were quickly made, they apologized for the wild greeting and began to calm down enough to explain further.

The couple were also guests at the campground and had retired several years before. This was their second stay at this spot and it turned out they were only three spaces away from our area. We explained that our rather standoffish attitude was because we automatically assumed it to be a joke people played on "newbies!"

We had of course, heard of Bigfoot, but had never given much thought to it, figuring it was most likely a hoax that was perpetuated by tourist bureaus and summer rental agencies and campgrounds. We reasoned that Roger and Linda had not wished to frighten anyone about the Sasquatch in case the people would think they were lying, and since anyone's chance of actually seeing this creature was evidently quite rare.

Well anyway, we enjoyed meeting Walt and Martha and visited often over our stay. After hearing their story, we thought it would be a good chance to see for ourselves, so with their directions, we headed up the trail. After about another six switchbacks, the trail turned left and began to wrap around the mountain.

United States Forest Service [Public domain]

By now, we were high enough up to see the beautiful scenery from a wide panorama. We could identify Oregon's Mount McLoughlin, even though there was a blue haze that covered the entire picture.

As we walked further on the trail, it dipped down, and instead of a wide ridge, we dropped gradually into a sort of large bowl. As we descended, our scenic view rapidly disappeared, and we soon found ourselves in this grassy meadow; the bottom of which was surrounded by both deciduous and very large pine trees.

Once we reached the bottom of the slope, the trail stretched out through the waist high grass, and there were many trails branching off in all directions. After another 50 yards, the trail once again began to descend slightly, and it was then that we saw the smallish lake where Walt and Martha said they saw the Bigfoot.

We began walking more slowly in anticipation, and Danice pulled her camera out of her vest pocket to be prepared. The trail had now steepened enough that Danice put the camera back in her pocket and both of us were very slowly descending the now treacherous slope. Having taken a single one of those collapsible aluminum hiking poles for each of us, our free hands seemed to be constantly acting as balance weights to keep us from falling and slipping on the now very gravelly slope.

Walt and Martha had not told us anything, but that it was steep; they should have said "dangerous," but maybe their experience was far advanced from that of us flatlanders. Finaly, the trail turned off at a sharp angle and leveled out.

We were now in a flat area, like the size of a soccer field, and the grasses were waist high. We could hear the crickets and the sounds so familiar from my childhood; a bullfrog. The unmistakable scent of a swampy area added to the certainty that water was nearby when the trail turned sharply, and there it was.

An almost picture-perfect small lake, covered for the most part with lilypads floating on the blue-black water, moving silently in the breeze, and small waves barely rippling the surface.

Suddenly, we were startled by the splash of a smallish, brown creature plunging into the dark water; it could have been an otter, muskrat, or perhaps even a beaver. Whatever it was, it caused us both to react in the most ruinous manner; Danice let out a semi-scream and I shouted, "Look!" Well, our reactions triggered another reaction which we didn't want at all!

United Geological Survey [Public domain]

Having stepped down the additional couple of feet to the water's edge, the surrounding grasses were now slightly above our eye level. That was when we heard the loud splashing and thrashing through the grasses to our right. We had startled some animal with our outbursts.

We quickly ran up to the path, and as we turned to where the sound was diminishing, we saw a large, ape-like creature retreating through the tall grasses. Within seconds, it had gained the heavily forested area, and then all was silent.

We reasoned that we hadn't seen the Sasquatch when we arrived, because it most likely had been fishing in the other end of the lake. Kicking ourselves for not having the precaution to always leave one to watch while the other one explores; we then realized that the camera was back in Danice's pocket, so we wouldn't have had a photo regardless.

Before we retraced our way back to camp, we searched the other side of the pond, and we did find one large depression on the edge of the dirt where it bordered the water, and it seemed almost double my size eleven boot, but it may have been smaller as it had filled with water almost to the top of the depression. We also found a large bullfrog that was still twitching slightly, but dead. I guess we ruined Bigfoot's lunch!

*Publishers note: Wikipedia describes Lake of the Woods this way: "**Lake of the Woods** is a natural lake near the crest of the Cascade Range in the Fremont–Winema National Forest in southern Oregon in the United States. The lake covers 1,146 acres (4.64 km). It was named by Oliver C. Applegate in 1870. Today, the Oregon Department of Fish and Wildlife manages the*

lake's fishery. The small unincorporated community of Lake of the Woods is located on the east shore of the lake. Lake of the Woods is one of southern Oregon's most popular outdoor recreation sites."

19 THE FOREST SASQUATCH'S HOME

Anonymous ~ Southern Oregon

I work for a private timber contractor, and in addition to the company's owned property, of which certain tracts come up for harvest on a regular schedule, we bid occasionally on parcels that hit maturity, but the owners have sold off their own equipment due to the uncertainty of the marketplace.

Times have been tough for our industry ever since the government began getting involved with trying to save endangered species, such as the spotted owl. So, as usually happens when Uncle Sam goes all out to save one species, it puts other ones in danger.

In this case, a lot of our fellow logging outfits threw in the towel, and now we have huge tracts that have been destroyed by forest fires, and the Feds haven't even allowed us to bid on the downed timber! What we have now are thousands of acres of forests that either fully or partially burned, and even worse, the timber that could have been salvaged has rotted and lies decaying across hundreds of square miles. In addition, all of the logging roads that once networked throughout this mountainous terrain are blocked with rotting logs. They too should have been salvaged!

So now, if we want to bid on a tract of timber, we incur the additional cost of clearing and rebuilding the hundreds of miles of roads just to make a marginal profit if any!

Yes, I am bitter, because at 62 years of age, I should be better off financially, but with the best years of my career spent barely surviving, I'll have to try making it up by just never retiring; thanks Uncle!

Anyway, now that I'm through venting, the owner of our company recently scored a very lucrative contract and our team who remained loyal to him over the tough years will be making very good wages. For a while anyway.

We spent the first month preparing for this job by first working in the early spring, which generally is not done, but we first had to clear the roads before the spring thaw and the forming of impassable, muddy roads. Bulldozers and chainsaws roared ten hours a day, but we got everything in this huge tract cleared, so when the roads became again passable, we were ready to "make bank!"

When we assaulted this multi-million dollar tract after roads became passable, in order to make the most time, the owners made a huge logging camp. Complete with trailers, a mess hall, an emergency medical tent with experienced technicians; this was like a small city amidst the enormous mountains.

We worked six days a week, and by the time winter chased us back out, we were feeling bodily aches, but "no pain" financially!

The only problem that we experienced were the annoyances of a group of Sasquatch! Being as how these animals were familiar to every one of us, including cooks and medics, we didn't enter into the initial shock of encountering these mountain apes. They in turn, were familiar with humans.

This was the first time that we had problems however. The reason being, we felt that because of our long hiatas, these critters probably felt that we were gone for good and the millions of acres had reverted back to them. After all, they were here first!

We immediately began having problems, even before our camp was established and staffed. I suppose it was a shock to the poor beasts who had always had to suffer the intrusion into their homes, but never before had any teams ever created an entire village in the midst of their forest homes.

The owners finally succumbed to hiring guards who worked in shifts with bright spotlights, and though we all lost some sleep on the occasions where the sounds of gunshots and shouts among the roving guards caused commotion, as tired as we were normally, nothing short of a helicopter landing could arouse most of us.

I am not at liberty to speak further on the subject due to the employment contract I signed; it has a privacy clause that prohibits all of us from discussing any and all happenings at our camp. Since there is a very substantial bonus due each of us at the end of our three year contract, I will say no more about the armed guards.

Publishers note: We accepted this story contrary to our desire to "verify for ourselves" the validity of those submitting their stories. In this case, we know the events surrounding the story, such as the spotted owl protection are true. Therefore, we agreed to print this submitter's story, but do not have verification of the personal events.

Two other loggers that we know and trust felt this information, "my possibly have really happened." We also have it on good authority from an outside source who was not part of the employment contract that Sasquatch killings did indeed take place!

The informant then told us that he intended to reveal the entire story and the huge "body count" to local authorities, and that was the last we ever saw of the man. We had reason to believe that this man was one of the participants in the shootings, so who knows what went on?

We had already moved out of Oregon when we received the final reports, but nothing was ever spoken of again, and the massive timber harvest ended the following year.

It is our hope that this Sasquatch killing did not actually take place, but if it truly did occur, it seems to have a solid curtain around everything involving any facet of this activity!

20 HIBERNATION OR MIGRATION

Gary Swanson

Hibernation is certainly a subject that may involve the Sasquatch. Many of our contributors have espoused the theory that they hibernate like bears. Scientists are pretty much removed from the question as to whether or not the Sasquatch hibernates, as they still have not settled the issue of whether or not they "officially" are accepting its existence in the first place.

We, as publishers of the stories submitted to us, try to be neutral on the subject out of fairness to our readers. We have however, had our own encounters that we have previously reported, questioning whether the Sasquatch hole up or move to warmer climes.

The hibernation theory has been occasionally touched on, but there are conflicting viewpoints on the subject. Such as, the theory that maybe, instead of hibernation that is imposed by nature due to the scarcity of a wintertime food source, that the Sasquatch has the ability to adjust its dietary needs to it having an available food source much higher in protein.

One dietary technician at a well-known hospital, suggests that perhaps, "The animal may adjust its metabolism to allow for a much greater intake of protein at winter's approach." An example she gives is for the animal to concentrate on an almost meat-only diet which is higher in protein, and therefore, unlike the black bear, it can avoid slowing down its metabolism by this method, thus avoiding the need to hibernate. This theory does make a lot of sense when one considers the fact that in states like Georgia and Florida, the black bear doesn't hibernate at all.

When analyzing this theory further, it would be hard to verify the proof, as during the heavy winter snows, one would find it difficult to inspect the possibility, as not many people would, or even could, venture far enough into the backcountry where these solitary beings are known to exist to find evidence of their existence, let alone their eating habits!

The occasional gut pile that one would expect from most carnivores perhaps would have been eaten immediately by a fox, wolf, weasel or the like. Also, to be in the right place when the kill is fresh would require armies of researchers.

We know of very few people who would be willing to don skis or snow shoes to tread silently into our winter wilderness to, on the off chance, perhaps find a fresh gut pile. Then how could one tie it in to a Sasquatch kill? As of yet, no Sasquatch enthusiast has volunteered!

Another way to look at this theory is that a bear's caloric intake is likely around 10 to 20 percent of animal protein, while according to what some of our sources say, the Bigfoot's diet may be much more dependent on meat than that.

A veterinarian who have previously reported on her personal confrontation and then her follow-up study of Sasquatch, has told us that she believes the animal's diet must be around 40 to 55 percent animal protein.

As publishers of these stories, we try to keep from inserting our personal opinions into the submissions from our growing list of contributors. The opinions that we form sometimes change from one theory to the next, based on the stories we receive. The hibernation theory for example, is one which seemed to be the most logical until a few months back when we were having dinner with a well-known explorer and anthropologist who espoused a theory that one of his colleagues had brought forth; and that was how similar the bear and the Sasquatch may really be.

The doctor felt that in early spring as the bears come out of hibernation, they roam quickly and at great distances to take advantage of the still preserved remains of animals that had frozen to death over the winter. He went on to mention that he was not convinced that the Sasquatch actually existed, but there was some evidence that the Sasquatch competed with the bear for these dead and frozen animals.

Our friend said he was doubtful of that theory, however, he mentioned reports of more than a few Native Americans, that he knew personally, from a nearby reservation who claimed to know of times that the Bigfoot had killed and eaten various animals that had just been born in the spring; so both sources add to the confusion.

Our opinion on the hibernation theory has shifted back to neutral; perhaps as it should be, as it is not our intent to prove or disprove any parts of the existence of this elusive animal. We do not have the qualifications to do more than speculate.

We have ourselves, witnessed several experiences that convinced us that the Sasquatch does exist, but as far as all the other specifics, we rely solely on our growing list of contributors to submit their experiences, and we welcome their stories and their resultant theories!

ABOUT THE AUTHORS

Gary Swanson had spent 40 years in the Pacific Northwest before he and Wendy moved to Grants Pass, Oregon, where they enjoyed hiking throughout the spring, summer and fall months with their dogs. In addition to their love of hiking, they also enjoy history. Southern Oregon is full of history of gold mining, logging and fishing along the wild and scenic Rogue River; so for them, it had been a great place to research history, explore the countryside and hike all at the same time.

Although they have relocated to sunny Southern Utah, they still enjoy hiking with their miniature schnauzers, and as more Sasquatch stories keep arriving, they have also begun to receive some very guarded and secretive information about an almost unbelievably evil creature called "Skinwalker." They are understandably leery of this research, because where Sasquatch is really interesting, the Skinwalker seems disgusting and dangerous. They received many stories of this very evil creature and published them in "Skinwalkers, Shapeshifters and Native American Curses", "The Last Skinwalker" and "We Survived Native American Witches, Curses and Skinwalkers".

If you have had a sighting or an encounter with Sasquatch, or even a Skinwalker, and would like your story published, the Swanson's welcome you to send your contact information, details of the encounter and any photos to swanliterary@gmail.com.

www.ingramcontent.com/pod-product-compliance
Lightning Source LLC
Chambersburg PA
CBHW070234180526
45158CB00001BA/496